Multiscale Modeling and Characterization Technique of Advanced Nanostructured Materials for Extreme Service

Multiscale Modeling and Characterization Technique of Advanced Nanostructured Materials for Extreme Service

Guest Editors

Hai Huang
Feida Chen

MDPI

Basel • Beijing • Wuhan • Barcelona • Belgrade • Novi Sad • Cluj • Manchester

Guest Editors

Hai Huang
Key Laboratory of Material Physics (Ministry of Education)
Zhengzhou University
Zhengzhou
China

Feida Chen
Department of Nuclear Science and Technology
Nanjing University of Aeronautics and Astronautics
Nanjing
China

Editorial Office
MDPI AG
Grosspeteranlage 5
4052 Basel, Switzerland

This is a reprint of the Special Issue, published open access by the journal *Nanomaterials* (ISSN 2079-4991), freely accessible at: https://www.mdpi.com/journal/nanomaterials/special_issues/RQK8I329L6.

For citation purposes, cite each article independently as indicated on the article page online and as indicated below:

Lastname, A.A.; Lastname, B.B. Article Title. *Journal Name* **Year**, *Volume Number*, Page Range.

ISBN 978-3-7258-7583-2 (Hbk)
ISBN 978-3-7258-7584-9 (PDF)
https://doi.org/10.3390/books978-3-7258-7584-9

Contents

About the Editors

Hai Huang

Hai Huang is an Associate Professor and Master's Supervisor at the Key Laboratory of Material Physics, Ministry of Education, School of Physics, Zhengzhou University, China. His interdisciplinary research focuses on nuclear technology and advanced materials, with expertise in radiation-resistant materials for nuclear reactors, structural materials for renewable energy systems, and the multi-scale computational modeling of materials. He earned his Ph.D. in Nuclear Technology and Materials Engineering from the Nanjing University of Aeronautics and Astronautics in 2019 and was a Visiting Scholar at the University of Michigan, Ann Arbor, in 2018. He has been recognized for his academic contributions through acceptances into programs such as the Henan Provincial Young Talent Support Program, Training Program for Young Backbone Teachers at Zhengzhou University, and Zhengzhou University Elite Doctoral Talent Program. He has led multiple research projects, including grants from the National Natural Science Foundation of China, Natural Science Foundation of Henan Province, and Key Projects for Science and Technology Development of Henan Province. His work has resulted in 60+ publications, 8 patents, and 10 software copyrights. He also actively contributes to academia as a Guest Editor for *Nanomaterials* and *Materials*, an Early Career Editorial Board Member of *Tungsten*, *Nanomaterials*, and *Copper Engineering*, and a reviewer for leading journals such as *Applied Surface Science* and the *Journal of Nuclear Materials*. His scientific pursuits continue to advance materials science for nuclear and renewable energy applications.

Feida Chen

Feida Chen is an Associate Professor in the Department of Nuclear Science and Technology at Nanjing University of Aeronautics and Astronautics. He earned his Ph.D. in Nuclear Technology from the Nanjing University of Aeronautics and Astronautics in 2016 and subsequently completed a postdoctoral fellowship at the GSI Helmholtz Centre for Heavy Ion Research in 2019. His research centers on the radiation effects of structural materials in nuclear engineering, with a specific focus on developing innovative techniques, such as laser shock peening, additive manufacturing, and pulsed electric currents to mitigate irradiation-induced defects and improve radiation resistance in critical metallic components for nuclear power systems, thereby supporting component repair and life extension. He also explores the irradiation modification of functional materials, including thermoelectric and electrically insulating materials.

Preface

The closure of our *Nanomaterials* Special Issue, "Multiscale Modeling and Characterization Technique of Advanced Nanostructured Materials for Extreme Service," does not mark an end but a new beginning. It is with great pleasure that we present this reprint, which captures the rich and diverse research contributions that defined this collection. Our primary motivation for compiling this work stemmed from the pressing need to understand and predict material behavior under extreme conditions—high radiation, intense thermal and mechanical stress, and corrosive environments—which are critical for advancements in nuclear, aerospace, and deep-space technologies.

This reprint has a clear scope: to showcase a powerful, interdisciplinary approach that integrates cutting-edge experimental characterization with state-of-the-art multiscale modeling. The collected articles brilliantly demonstrate this synergy, covering topics from the first-principles calculation of defect evolution and molecular dynamics simulations of irradiation damage to the development of novel composites and machine learning-driven materials design for applications such as radiation shielding and nuclear fuel management. The aim is to provide a definitive snapshot of the current frontier in this field and to inspire future breakthroughs.

We are grateful to all of the pioneering researchers who contributed their outstanding work to this Special Issue and reprint. Their collective expertise, spanning materials science, physics, chemistry, and engineering, is the true strength of this volume. This reprint is addressed to a broad audience of scientists, engineers, and graduate students who are dedicated to pushing the limits of material performance. We extend our appreciation to the MDPI Editorial Team for their invaluable support and professionalism throughout this process. Finally, we thank the dedicated peer reviewers whose critical insights ensured the high quality of this published science. We hope this compilation serves as a valuable resource and a catalyst for continued innovation.

Hai Huang and Feida Chen
Guest Editors

nanomaterials

Article

Enhanced Rashba Effect and Optical Absorption in 2D Janus $XMoYZ_2$ (X = S/Se/Te; Y = Si/Ge; Z = N/P): A First-Principles Study

Xiaochuan Liu, Meng Li *, Ningru Shang, Peng Guo, Hongyue Song, Bin Zhao, Lin Li and Jianjun Wang *

Zhengzhou Key Laboratory of Low-Dimensional Quantum Materials and Devices, School of Physics and Optoelectronic Engineering, Zhongyuan University of Technology, Zhengzhou 450007, China
* Correspondence: limeng@zut.edu.cn (M.L.); jjwang@zut.edu.cn (J.W.)

Abstract

To overcome the physical constraints during the miniaturization of conventional semiconductor devices, spintronics is playing an increasingly prominent role. The Rashba effect, characterized by spin–momentum locking, has emerged as a promising solution to address challenges. Two-dimensional (2D) Janus transition metal dichalcogenides (TMDCs) break spatial inversion symmetry, creating favorable conditions for the Rashba effect. Based on first-principles calculations, 2D Janus materials $XMoYZ_2$ (X = S/Se/Te; Y = Si/Ge; Z = N/P) were investigated, with strain, external electric field and charge doping employed to modulate the Rashba effect. The strain results reveal that the Rashba constants of $XMoYZ_2$ increase significantly with compressive strain. Specifically, after applying uniaxial strain, the Rashba constant of $TeMoSiP_2$ is enhanced to ~2.2 times its initial value. Compressive strain reduces atomic spacing, enhances orbital overlap, and increases spin–orbit coupling (SOC) strength. All the $TeMoYZ_2$ materials exhibit significant anisotropy under uniaxial strain, which is favorable for spin-oriented transport. $SeMoGeP_2$ shows an almost linear Rashba constant–electric field correlation, while $TeMoGeP_2$ and $TeMoSiP_2$ show non-monotonic variation. The Rashba constant of $TeMoSiP_2$ can be enhanced to ~2.7 times its intrinsic value under either positive or negative applied electric fields. Charge doping induces negligible changes in the SOC effect. Finally, the optical absorption properties of $TeMoGeP_2$, $TeMoSiN_2$, and $TeMoSiP_2$ were investigated. This study clarifies the mechanism underlying the enhancement of Rashba constants in $XMoYZ_2$ materials, enriching the research landscape of spintronics.

Keywords: spintronics; Rashba effect; 2D Janus $XMoYZ_2$; optical absorption; first-principles

1. Introduction

As Moore's Law approaches physical limit, researchers have leveraged spin transport to mitigate quantum tunneling induced by charge-based transport. With the advancement of spintronics, the Rashba effect [1–6], arising from the effective magnetic field generated by moving electrons, has emerged as a research hotspot. It establishes a correlation between electron momentum and spin orientation via a specific energy term [7,8]. The Rashba effect can be manipulated during spin transport processes, achieving spin filtering through valley polarization [9,10]. A strong Rashba effect implies that a weak external field can achieve modulation of spin directions, reducing energy consumption [11]. Based with spin–

 https://doi.org/10.3390/nano16060358

momentum locking, the Rashba effect enables novel architectures to overcome bottlenecks, paving the way for high-performance devices [12].

Two-dimensional materials have become an ideal platform for the Rashba effect. However, as a zero-bandgap semiconductor, graphene is unfavorable for discrete information storage, and the stability of black phosphorus and the carrier mobility of traditional 2D materials pose challenges; although strong Rashba effects can been observed in several bulk materials, it remains unclear whether such effects can persist at the atomic scale [13]. Recent studies have demonstrated that 2D Janus materials exhibit excellent Rashba effects and spin tunability, attributed to their spatial inversion asymmetry [14–16], and transition metal dichalcogenides (TMDCs) have emerged as a focus [17,18]. These materials possess suitable bandgaps and intrinsic out-of-plane electric fields induced by different vertically aligned atoms [19–23]. Reports have revealed that such intrinsic fields exhibit superior efficacy in enhancing the Rashba effect compared to externally electric fields.

The Rashba effect typically occurs in confined two-dimensional electron gas systems. The Bychkov–Rashba Hamiltonian can be described by the $k \cdot p$ model [24].

$$H_R = \alpha(\sigma \times p) \cdot \hat{z} = \alpha(\sigma_x p_y - \sigma_y p_x) \tag{1}$$

where α denotes the Rashba constant, $p = \begin{pmatrix} p_x \\ p_y \end{pmatrix}$ represents the momentum operator, $\hat{z}$ is the out-of-plane unit normal vector, and the spin operator $\sigma = (\sigma_x, \sigma_y, \sigma_z)$ corresponds to the Pauli spin matrices.

The Hamiltonian of the system can be expressed by the following formula.

$$H = \frac{p^2}{2m^*} + V_0(r) + \frac{e\hbar^2 E_z}{4m^{*2}c^2}(\sigma_x p_y - \sigma_y p_x) \tag{2}$$

where the first term represents the kinetic energy term (m^*: effective mass), the middle term denotes the external electric potential, the last term corresponds to the Rashba SOC Hamiltonian, and E_z indicates the perpendicular electrostatic potential gradient.

Let

$$\alpha_R = \frac{e\hbar^2 E_z}{4m^{*2}c^2} \tag{3}$$

The eigenvalues of the Rashba effect are derived as

$$E_{R\pm} = \frac{\hbar^2 k^2}{2m^*} \pm \alpha_R k \tag{4}$$

The Rashba constant α_R is given by the quotient of the difference in energy and difference in momentum from the splitting point to the VBM or the CBM

$$\alpha_R = 2\frac{E_R}{k_R} \tag{5}$$

In this paper, Rashba splitting occurs in the valence band; E_R and k_R are defined as

$$E_R = |E_\Gamma - E_{\max}| \tag{6}$$

$$k_R = |k_\Gamma - k_{\max}| \tag{7}$$

where $E_{\max}$ and E_Γ represent the maximum energy along two directions around and at the Γ point, respectively; $k_{\max}$ and k_Γ are the momentum values corresponding to $E_{\max}$ and E_Γ.

For uniaxial strain, the lattice constants along the x-axis follow

 https://doi.org/10.3390/nano16060358

$$a = a_0(1 + \varepsilon_x) \tag{8}$$

The other lattice constants follow

$$b_x = b_{0x}(1 + \varepsilon_x),\ b_y = b_{0y}(1 + \varepsilon_x) \tag{9}$$

Freitas et al. pointed out that the density of states, bandgap type and size, orbital contributions, and topological insulating properties of XBi and XBi_3 all undergo changes when SOC was considered, demonstrating that SOC can influence the topological properties [25]. Szczęśniak proposed the concept of metal induced gap states, transforming the insulating gap into metal like through altering the electronic distribution, governing Fermi level pinning, and mediating valley–spin polarized charge injection and transport, predicting that such states could mediate valley-resolved and spin-resolved charge transport near the electrode/MX_2 interface, thereby resulting in valley–spin filtering [10]. Compared with reports on the strong Rashba effect in BiTeI monolayers, $XMoYZ_2$ exhibits significant anisotropy, which can generate spin currents in a directional manner. This is an advantage that is difficult to achieve with BiTeI. Even though α_R has not reached a record value, its anisotropy and controllability provide a new dimension for spin devices [26,27]. Wei et al. screened 26 Janus $MAA'Z_xZ'_{(4-x)}$ layers and found that the Rashba effect in these materials is dominated by Se/Te-p_z orbitals [2]. The α_R reached 0.89 eV·Å, revealing that the intensity of the Rashba effect is closely correlated with the built-in electric field and the SOC strength. Farooq et al. showed that spin–momentum locking can be switched by a gate voltage in AB′ stacked InSe [28]. Rezavand et al. investigated 15 Janus TMDC stacking structures and found that Rashba splitting only exists in adjacent stacks with different chalcogen atoms; this effect is the most pronounced in AA-XY and A′B-XY stacking configurations, and can be significantly enhanced by an out-of-plane electric field or appropriate in-plane strain [29,30]. Hu et al. applied a high electric field of 0.5 V/Å to WSeTe, increasing its Rashba constant by 0.31 eV·Å, which confirmed the feasibility of two-dimensional materials under high electric fields [31]. These studies have shown that the strength of Rashba spin splitting can be significantly tuned by strain [32]. Su et al. studied the Rashba effect of SnS and SnSe monolayers, and found that applying strain along the ZZ, AC and biaxial directions could not only induce a bandgap transformation, but also effectively modulate the momentum shift and splitting strength [33]. Specifically, the α_R of the SnS monolayer reached 0.76 eV·Å under 6% ZZ tensile strain while that of SnSe reached 1.33 eV·Å under 2% biaxial tensile strain. In the experiment, Sibatov et al. proposed that $XMoSiN_2$ can be achieved by substituting chalcogen elements for SiN_2 on one side of $MoSi_2N_4$ [27], and $MoSi_2N_4$ has been successfully prepared experimentally by chemical vapor deposition (CVD) [34]. Strain can be realized through epitaxial growth on substrates with lattice or thermal expansion mismatch and direct stretching or compressing of flexible substrates [35].

The $XMoYZ_2$ (X = S/Se/Te, Y = Si/Ge, Z = N/P) series is selected to investigate the structure–property relationships. Chalcogen X is used to tune spin–orbit coupling strength, group-IV Y modulates the electronic structure near the Fermi level, and pnictogen Z (N/P) provides complementary bonding and valence orbital features. This allows us to clarify their individual effects on the electronic and spin-related properties. Zamanian et al. reported the intrinsic Rashba constants of $XMoYZ_2$ but the modulations for enhancing Rashba constants were not elaborated [25]. Therefore, the strain, electric field and charge doping were adopted to improve the Rashba constants.

2. Computational Methods

First-principles calculations were performed, with the Perdew–Burke–Ernzerhof (PBE) exchange correlation [36,37]. The Projector Augmented Wave (PAW) was adopted [38]. The cutoff energy of 500 eV was set for all calculations. The convergence criterion for elec-

https://doi.org/10.3390/nano16060358

tronic iterations was fixed at 10^{-8} eV while the ionic relaxation was 10^{-4} eV/Å. Gaussian broadening with a width of 0.05 eV was used. A 15 × 15 × 1 Monkhorst–Pack k-mesh was employed for structural optimization and optical absorption calculations. Too dense k-point samples would result in the same energy corresponding to different k-points; therefore, selecting 301 samples is sufficient enough to ensure the accuracy of Δk. The out-of-plane lattice constant was extended to 30 Å to ensure an approximately 15 Å vacuum layer. Bader charge analysis was incorporated to quantify the charge transfer [39]. The SOC was considered in all calculations.

Among all the modulations, the atoms' relaxation was first calculated for all sampling points. When applying strain, we fixed the calculated lattice constants. When applying electric fields, the dipole correction was activated and its position was set at the center; the electric fields were set to gradually increase along the z axis. Valence electron numbers ± ne means that 1 e per unit cell is equivalent to about 0.0082 e per $Å^2$. (For instance, taking the $TeMoSiP_2$ lattice constant as an example, when doping with 0.1 e, it means that 10 unit-cells are doped with 1 e, and the charge density is $\frac{1}{3.493^2 \times 10} \approx 0.0082\ (e/Å^2)$).

Phonon dispersions were calculated to evaluate the dynamical stability. The force constants were obtained via Density Functional Perturbation Theory (DFPT) [40,41]. The unit cell was expanded to 2 × 2 × 1, and a 5 × 5 × 1 k-mesh was used.

3. Results and Discussion

3.1. Structure

As illustrated in Figure 1a–c, $XMoYZ_2$ structures crystallize in the hexagonal crystal system. Based on this stacking structure, twelve monolayers were constructed, namely, $SeMoGeN_2$, $SeMoGeP_2$, $SeMoSiN_2$, $SeMoSiP_2$, $SMoGeN_2$, $SMoGeP_2$, $SMoSiN_2$, $SMoSiP_2$, $TeMoGeN_2$, $TeMoGeP_2$, $TeMoSiN_2$ and $TeMoSiP_2$. The chalcogenides are located at the bottom layer, the transition metal Mo at the fourth layer, N/P at the third and first layers and Si/Ge occupy the second layer. The $XMoYZ_2$ structures belong to the C_3 Schönflies point group. Rotations around the axis passing through Mo and perpendicular to the monolayer plane (via $C_3{}^1$ and $C_3{}^2$ operations) can map the $XMoYZ_2$ structures onto themselves. The Janus structures, induced by the asymmetric vertical layer distribution, break the C_3h point group symmetry, which directly results in the absence of spatial inversion symmetry. The lattice constants and other critical data are summarized in Table 1, with reference values provided for comparison.

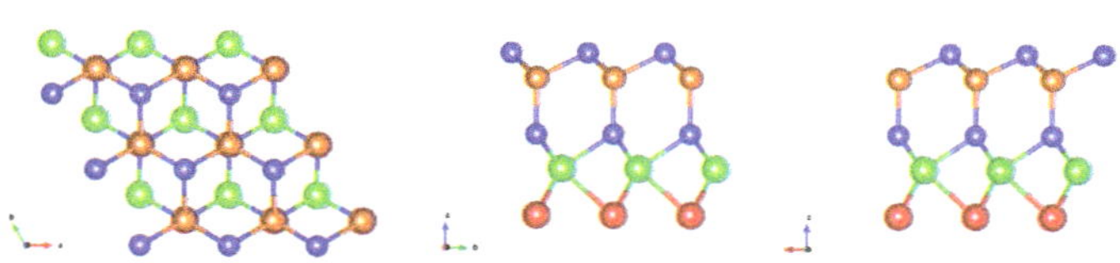

(a) (b) (c)

Figure 1. The structure of $XMoYZ_2$. (**a**) Top view and (**b**), (**c**) side views; the blue, orange, green and red balls represent Z, Y, Mo, and X elements, respectively.

Table 1. The lattice constants and bond lengths of $XMoYZ_2$; Z_1 and Z_2 represent the middle layers and the outside layers, respectively.

	α(Å), a = b	Ref. [22]	$l_{X\text{-}Mo}$ (Å)	$l_{Mo\text{-}Z1}$ (Å)	$l_{Z1\text{-}Y}$ (Å)	$l_{Y\text{-}Z2}$ (Å)
$SeMoGeN_2$	3.103	3.100	2.511	2.137	1.882	1.881
$SeMoGeP_2$	3.478	3.471	2.585	2.449	2.325	2.300
$SeMoSiN_2$	3.013	3.009	2.499	2.115	1.752	1.802
$SeMoSiP_2$	3.428	3.421	2.574	2.439	2.239	2.236

 https://doi.org/10.3390/nano16060358

Table 1. *Cont.*

	α(Å), a = b	Ref. [22]	$l_{X\text{-}Mo}$ (Å)	$l_{Mo\text{-}Z1}$ (Å)	$l_{Z1\text{-}Y}$ (Å)	$l_{Y\text{-}Z2}$ (Å)
$SMoGeN_2$	3.067	3.067	2.390	2.132	1.878	1.867
$SMoGeP_2$	3.436	3.434	2.473	2.444	2.324	2.288
$SMoSiN_2$	2.980	2.980	2.374	2.109	1.751	1.786
$SMoSiP_2$	3.386	3.385	2.461	2.434	2.239	2.223
$TeMoGeN_2$	3.169	3.169	2.696	2.150	1.889	1.908
$TeMoGeP_2$	3.545	3.542	2.750	2.460	2.328	2.320
$TeMoSiN_2$	3.077	3.077	2.690	2.130	1.755	1.833
$TeMoSiP_2$	3.493	3.493	2.743	2.448	2.240	2.256

3.2. Stability

Figure 2 displays the phonon dispersion of $XMoYZ_2$, which consists of three acoustic branches and twelve optical branches. Small imaginary frequencies are observed near the Γ point. Such weak imaginary frequencies are often not indicative of instability but rather arise from computational artifacts or finite size, which means that their frequencies are easily affected by noise in force calculations, insufficient relaxation, or inadequate sampling of the Brillouin zone, all of which can introduce small imaginary frequencies, particularly for low-energy flexural phonons [42,43].

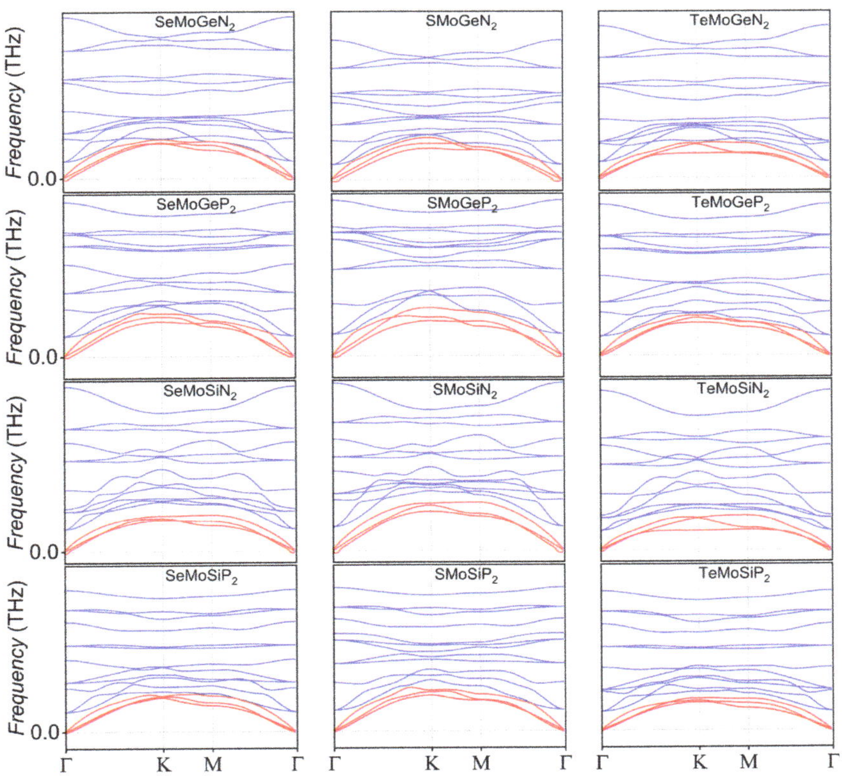

Figure 2. The phonon dispersion of $XMoYZ_2$. The red lines represent acoustic branches, and the blue lines denote optical branches.

https://doi.org/10.3390/nano16060358

3.3. Potential Energy and Charge Analysis

Based on Figure 3a and Table 2, $TeMoSiN_2$ exhibits the largest potential difference, which corresponds to the other studies [22]. The high electronegativity of N atoms endows them with a stronger electron-attracting capability than P atoms, leading to a larger electronegativity difference between N and Y/Mo atoms. This electronegativity discrepancy results in remarkable charge transfer in $XMoYN_2$ materials. For instance, for $TeMoSiN_2$, the charge transfer of the two N layers reaches −1.613 e and −1.842 e, respectively, which contributes to the stronger vertical intrinsic electric field. This electric field may exhibit high coupling efficiency with the Mo-$d_z{}^2$ orbital, thereby enhancing the Rashba effect. In contrast, the electronegativity difference between P atoms and other constituent elements is relatively small, which consequently weakens the Rashba effect. The electronegativity of Te and Mo atoms is nearly equal, which makes the electron cloud of the Mo-$d_z{}^2$ orbital less likely to be attracted by the Te layer; instead, it tends to hybridize with the Z_2-p_z and Y-p_z orbitals. This result may enhance the SOC strength, rendering the Rashba constants of $TeMoYZ_2$ materials more sensitive to the modulation. In contrast, the electronegativity difference between S/Se and Mo atoms may induce a shift in the Mo-$d_z{}^2$ orbital electron cloud toward the X layer, which may impair hybridization efficiency with the p_z orbitals of the top layers. This further leads to weakened SOC, resulting in the Rashba effect and modulation potential of $SMoYZ_2$ or $SeMoYZ_2$ being inferior to those of $TeMoYZ_2$ counterparts.

Table 2. The potential difference Δϕ, charge accumulation and depletion per layer of $XMoYZ_2$.

	Δφ	Se/S/Te	Mo	N/P	Ge/Si	N/P
$SeMoGeN_2$	0.232	−0.281	0.988	−1.191	1.861	−1.377
$SeMoGeP_2$	0.018	−0.556	0.934	−0.048	−0.279	−0.050
$SeMoSiN_2$	0.194	−0.464	1.299	−1.680	2.918	−2.073
$SeMoSiP_2$	0	−0.161	0.209	−0.012	0.748	−0.783
$SMoGeN_2$	0.027	−0.522	1.244	−1.215	1.893	−1.400
$SMoGeP_2$	−0.156	−0.292	0.200	0.380	−0.279	−0.008
$SMoSiN_2$	0	−0.537	1.317	−1.647	3.000	−2.133
$SMoSiP_2$	−0.224	−0.225	0.001	0.233	0.822	−0.831
$TeMoGeN_2$	0.460	−0.385	1.254	−1.347	1.825	−1.347
$TeMoGeP_2$	0.176	0.221	−0.321	0.248	−0.330	0.183
$TeMoSiN_2$	0.492	−0.167	0.885	−1.613	2.737	−1.842
$TeMoSiP_2$	0.182	−0.147	0.211	−0.494	0.999	−0.569

Figure 3b,c reveal the intrinsic electric field inside the monolayers. Specifically, for $TeMoSiP_2$, the Te layer loses a net charge of 0.147 e, while the Mo layer and its adjacent P layer accumulate and lose charges of −0.211 e and 0.494 e, respectively. The Si layer gains a net charge of 0.999 e, and the outermost P layer loses a net charge of 0.569 e. Notably, the charge transfer of $SMoGeP_2$ and $TeMoGeP_2$ is opposite to that of most other monolayers. In case of $TeMoGeP_2$, for example, the Te layer loses a net charge of 0.221 e, while the Mo layer gains a net charge of 0.321 e; the P_1, Ge, and P_2 layers undergo charge changes of 0.248 e, 0.330 e, and 0.183 e, respectively. The charge accumulation and depletion for each layer are summarized in Table 2.

https://doi.org/10.3390/nano16060358

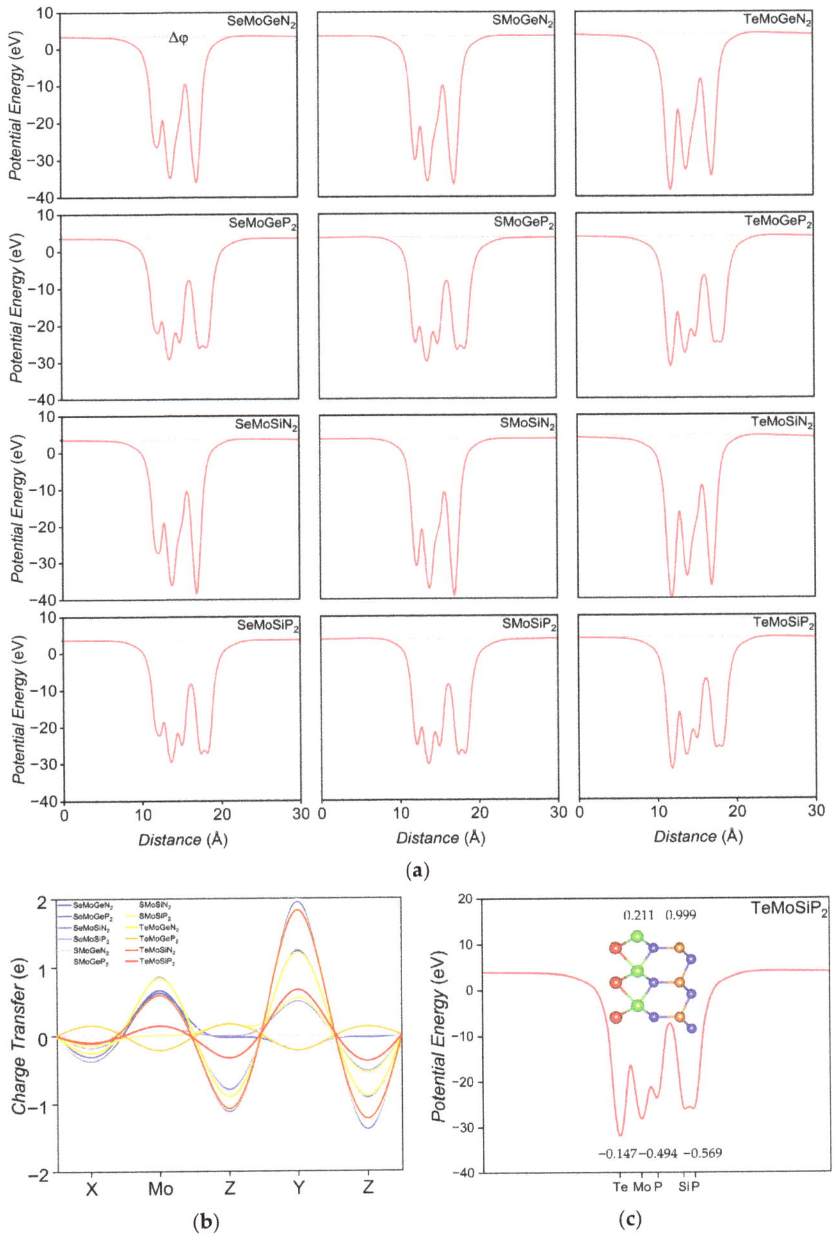

Figure 3. (**a**) Potential energy profile of $XMoYZ_2$ as a function of distance, with horizontal lines representing the potential difference Δϕ. (**b**) Charge transfer characteristics of $XMoYZ_2$; positive values denote charge depletion, while negative values signify charge accumulation. (**c**) Potential energy profile and charge transfer of $TeMoSiP_2$.

3.4. Spin Texture

Taking $TeMoSiP_2$ as an example, in Figure 4a, the degeneracy is lifted in the two energy surfaces where Rashba splitting occurs. Around the Γ point, the spin orientation is

https://doi.org/10.3390/nano16060358

clockwise on the upper energy surface but counterclockwise on the lower one, displaying opposite spin chirality. Away from the Γ point, the spin direction begins to develop outward and also exhibits the opposite spin chirality, which confirms that it is Rashba spin splitting.

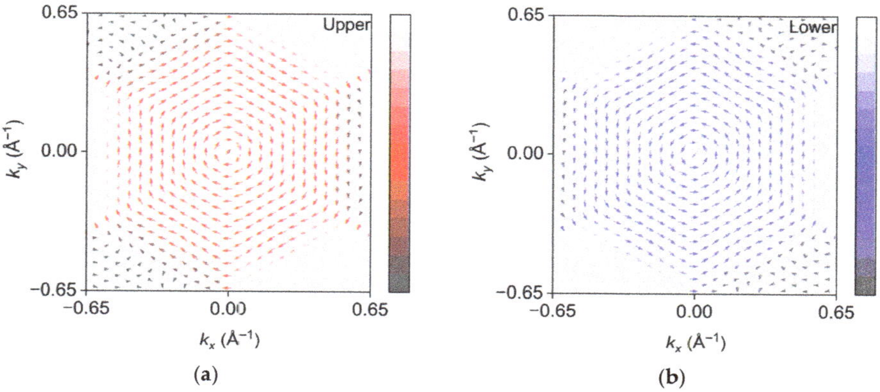

Figure 4. Spin textures of $TeMoSiP_2$; the arrow directions indicate the spin directions. Red represents in-plane spin-up, blue represents in-plane spin-down, and the arrows lengths and depths represent the in-plane spin component magnitude. White represents out-of-plane spin-up, and dark gray represents out-of-plane spin-down. (**a**) Upper energy surface and (**b**) lower energy surface.

3.5. Electronic Structure

As illustrated in Figure 5, when SOC is considered, the degeneracy of the electronic structures vanishes. Among the $XMoYZ_2$ monolayers, $SeMoSiP_2$, $SMoSiN_2$, and $TeMoSiP_2$ exhibit direct bandgaps at the K point. Except for $XMoGeN_2$, which shows Mexican hat-like splitting around the Γ point, all other monolayers display typical Rashba splitting. Specifically, the Rashba splitting of $SeMoGeP_2$, $SeMoSiN_2$, $SMoGeP_2$, $SMoSiP_2$, and $TeMoGeP_2$ is observed at the VBM. Relevant data are summarized in Table 3. When electrons or holes accumulate at the CBM and the VBM exhibits Rashba splitting, this can provide favorable conditions for spin polarization, which shows great promise in spintronic devices. The Mo-$d_z{}^2$ orbitals make significant contributions to Rashba splitting at the Γ point in almost all monolayers. A plausible reason is that the Mo-$d_z{}^2$ orbitals extend along the z-direction, while the electron clouds of elements such as N, P, and Se are also asymmetrically distributed along the z-direction, generating internal electric fields. The Mo-$d_z{}^2$ orbitals couple with these electric fields, enabling the SOC effect to be maximized and thus dominating spin splitting. In some monolayers, the Mo-$d_z{}^2$ orbitals match the p_z orbitals, forming hybrid orbitals; consequently, Rashba splitting is mainly composed of the hybridization between Mo-$d_z{}^2$ and p_z orbitals.

Table 3. Bandgap types, locations of Rashba splitting and bandgaps with SOC (E_g).

	VBM	CBM	Type	Location	E_g (PBE + SOC)
$SeMoGeN_2$	Γ	K	I		1.448
$SeMoGeP_2$	Γ	K	I	VBM	0.791
$SeMoSiN_2$	Γ	Γ→K	I	VBM	1.774
$SeMoSiP_2$	K	K	D		0.912
$SMoGeN_2$	Γ	K	I		1.512
$SMoGeP_2$	Γ	K	I	VBM	0.510
$SMoSiN_2$	K	K	D		2.081
$SMoSiP_2$	Γ	K	I	VBM	0.803

https://doi.org/10.3390/nano16060358

Table 3. *Cont.*

	VBM	CBM	Type	Location	E_g (PBE + SOC)
$TeMoGeN_2$	Γ	K	I		1.022
$TeMoGeP_2$	Γ	K	I	VBM	0.761
$TeMoSiN_2$	M	Γ→K	I		1.077
$TeMoSiP_2$	K	K	D		0.914

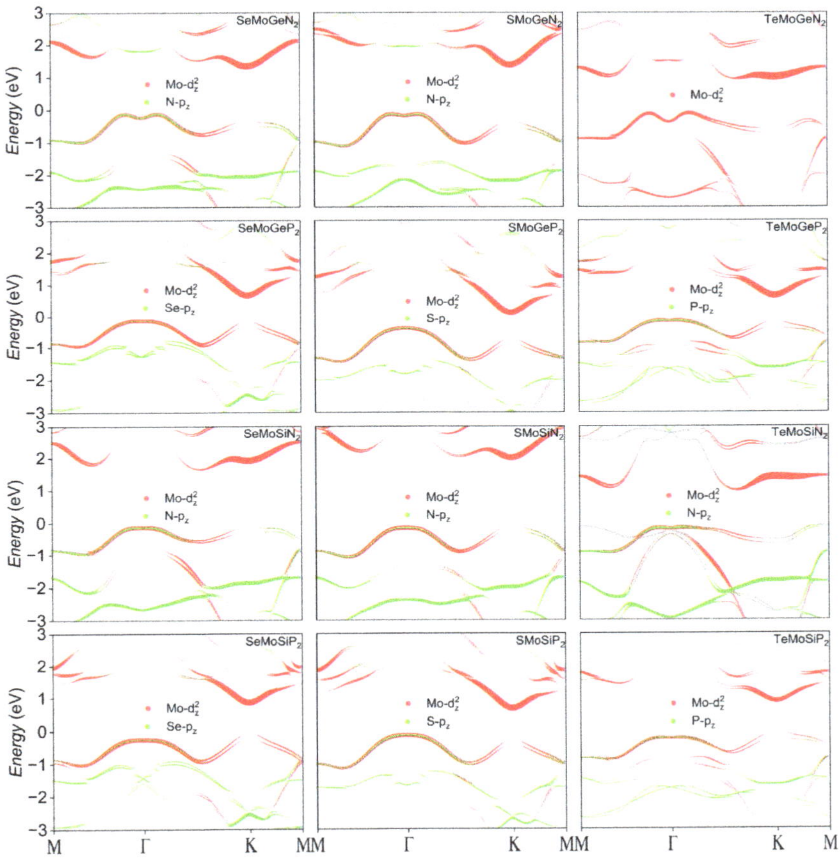

Figure 5. Electronic structures of $XMoYZ_2$ monolayers with SOC considered; orbitals that make significant contributions to Rashba splitting are marked.

3.6. Rashba Modulation

In the subsequent modulations, all structures satisfied the force convergence criteria and were confirmed to be stable.

3.6.1. Strain

$TeMoYZ_2$ was selected as the research object due to its excellent performance. Combining Figures 5 and 6a, under uniaxial strain, the Rashba constant of $TeMoSiP_2$ increases rapidly with the enhancement of compressive strain, with an intrinsic Rashba constant of 0.351 eV·Å. At this stage, the Mo-$d_z{}^2$ orbital is the main contributor, accompanied by the participation of a small amount of P-p_z and Si-p_z orbitals. Subsequently, upon applying −4% compressive strain, the Rashba constants along the Γ→M and Γ→K directions increase to 0.775 eV·Å and 0.553 eV·Å, respectively, which are 2.21 and 1.58 times the intrinsic value; the proportion of Mo-$d_z{}^2$ is 39.6% and that of P-p_x is 0.6%. From Table 4, the wave vector k where the peaks are located is shifting away from Γ, but ΔE increases faster than Δk. However, with the increase in compression, the Rashba constant shows a downward trend, which is attributed to the upward shift in the deeper energy levels in the valence band dominated by P-px orbitals; the proportion of Mo-$d_z{}^2$ is 17.9% and that of P-p_x is 17.9%. As illustrated in Figure 7a,b, these energy levels subsequently couple with Mo-$d_z{}^2$ orbitals and engage in Rashba splitting.

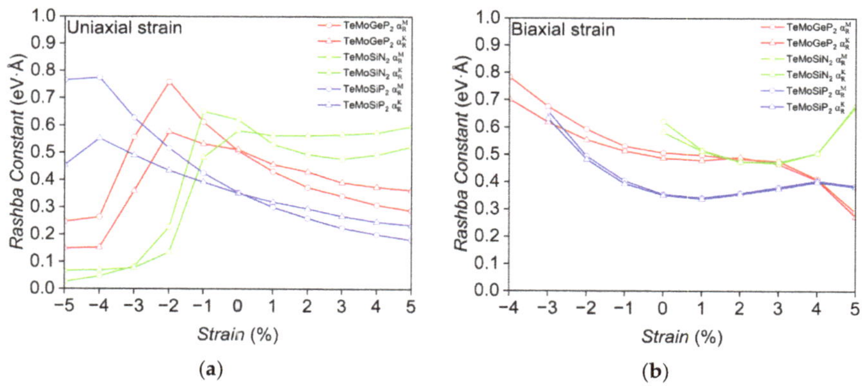

Figure 6. Variation in Rashba constants for $TeMoYZ_2$ under different strain types: (**a**) uniaxial strain and (**b**) biaxial strain. The red, green, and blue lines represent $TeMoGeP_2$, $TeMoSiN_2$, and $TeMoSiP_2$, respectively. Circular lines correspond to the Γ→M direction, while triangular lines correspond to the Γ→K direction.

Table 4. The variation in ΔE and Δk with the two kinds of strains of $TeMoYZ_2$.

		Uniaxial Strain (%)				Biaxial Strain (%)			
		$\Delta E_{\Gamma\to M}$	$\Delta k_{\Gamma\to M}$	$\Delta E_{\Gamma\to K}$	$\Delta k_{\Gamma\to K}$	$\Delta E_{\Gamma\to M}$	$\Delta k_{\Gamma\to M}$	$\Delta E_{\Gamma\to K}$	$\Delta k_{\Gamma\to K}$
$TeMoGeP_2$	−5	0.00044	0.00355	0.00030	0.00399				
	−4	0.00093	0.00704	0.00061	0.00796	0.06253	0.15989	0.07630	0.21745
	−3	0.00487	0.01746	0.00285	0.01587	0.04990	0.14769	0.05651	0.18272
	−2	0.03415	0.09004	0.02741	0.09502	0.03818	0.12878	0.04120	0.14871
	−1	0.03054	0.09967	0.02836	0.10662	0.03016	0.11370	0.03165	0.12333
	0	0.02505	0.09892	0.02719	0.10634	0.02505	0.09892	0.02590	0.10634
	1	0.02049	0.09500	0.02336	0.10215	0.02185	0.08781	0.02241	0.09359
	2	0.01694	0.09075	0.02108	0.09799	0.01944	0.08026	0.01987	0.08109
	3	0.01424	0.08342	0.01910	0.09776	0.01656	0.06955	0.01688	0.07265
	4	0.01228	0.07952	0.01751	0.09363	0.01212	0.05903	0.01232	0.06060
	5	0.01091	0.07567	0.01625	0.08953	0.00658	0.04548	0.00667	0.04877
$TeMoSiN_2$	−5	0.00005	0.00408	0.00031	0.00920				
	−4	0.00010	0.00405	0.00032	0.00917				
	−3	0.00050	0.01206	0.00054	0.01372				
	−2	0.00544	0.04788	0.00278	0.04105				
	−1	0.04255	0.13066	0.04613	0.19107				
	0	0.04271	0.13754	0.04883	0.16790	0.04271	0.13754	0.04883	0.16790
	1	0.03727	0.14043	0.04587	0.16295	0.03716	0.14397	0.04032	0.15725
	2	0.03441	0.13940	0.04452	0.15804	0.03470	0.14640	0.03683	0.15571
	3	0.03398	0.14224	0.04470	0.15768	0.03571	0.15261	0.03751	0.15860
	4	0.03576	0.14505	0.04640	0.16181	0.04005	0.15871	0.04190	0.16580
	5	0.03962	0.15162	0.04958	0.16593	0.05822	0.17216	0.06041	0.17719

 https://doi.org/10.3390/nano16060358

Table 4. *Cont.*

		Uniaxial Strain (%)				Biaxial Strain (%)			
		$\Delta E_{\Gamma\to M}$	$\Delta k_{\Gamma\to M}$	$\Delta E_{\Gamma\to K}$	$\Delta k_{\Gamma\to K}$	$\Delta E_{\Gamma\to M}$	$\Delta k_{\Gamma\to M}$	$\Delta E_{\Gamma\to K}$	$\Delta k_{\Gamma\to K}$
	−5	0.06197	0.16197	0.04332	0.19000				
	−4	0.06223	0.16067	0.04805	0.17400				
	−3	0.04784	0.15232	0.03856	0.15700	0.05053	0.15346	0.05860	0.18545
	−2	0.03634	0.14060	0.03145	0.14500	0.03582	0.14483	0.03938	0.16315
	−1	0.02745	0.12906	0.02596	0.13225	0.02611	0.12938	0.02777	0.14132
$TeMoSiP_2$	0	0.02087	0.11770	0.02178	0.12392	0.02087	0.11770	0.02178	0.12392
	1	0.01600	0.10652	0.01848	0.11564	0.01874	0.10968	0.01938	0.11478
	2	0.01243	0.09551	0.01585	0.10740	0.01885	0.10521	0.01943	0.10973
	3	0.00990	0.08805	0.01382	0.10318	0.02045	0.10756	0.02110	0.11254
	4	0.00813	0.08069	0.01226	0.09899	0.02219	0.10984	0.02301	0.11531
	5	0.00697	0.07679	0.01114	0.09482	0.02226	0.11539	0.02326	0.12182

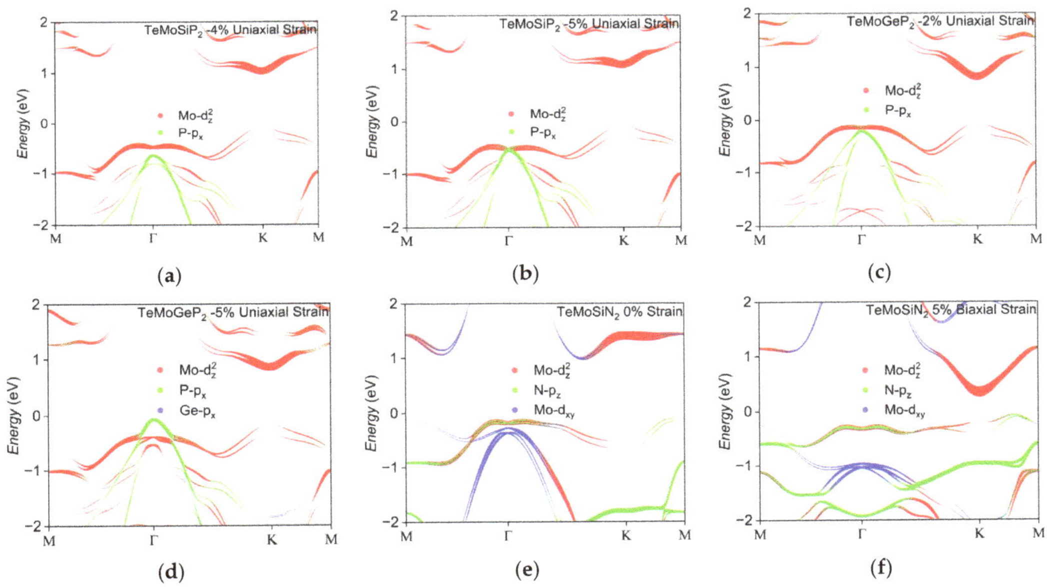

Figure 7. The electronic structures trend of $TeMoSiP_2$, $TeMoGeP_2$ and $TeMoSiN_2$ under different kinds and degrees of strain. (**a**,**b**) $TeMoSiP_2$ under −4% and −5% uniaxial strain, (**c**,**d**) $TeMoGeP_2$ under −2% and −5% uniaxial strain and (**e**,**f**) $TeMoSiN_2$ under 0% and 5% biaxial strain.

Similarly, $TeMoGeP_2$ also has an intrinsic Rashba constant of approximately 0.511 eV·Å, which is slightly higher than that of $TeMoSiP_2$. For the initial Rashba effect, the Mo-$d_z{}^2$ orbital is the main contributor, followed by the P-p_z orbital. As shown in Figure 6a, under a compressive strain of −2%, the Rashba constants reach their maximum values: 0.759 eV·Å along the Γ→M direction and 0.577 eV·Å along the Γ→K direction. This is because ΔE increases continuously while Δk decreases compared to that under −1% or 0% strain, leading to an increase in the Rashba constant. However, after exceeding this threshold strain, a downward trend also occurs. As shown in Figure 7c,d, when the compressive strain increases from −2% to −5%, the Rashba splitting region originally dominated by the Mo-$d_z{}^2$ orbital is completely replaced by the deeper-lying P-p_x orbitals; the proportions of Mo-$d_z{}^2$ and P-p_x change from 33% and 4.8% to 0.2% and 36.6%, respectively, showing a

https://doi.org/10.3390/nano16060358

prominent profile, resulting in almost no typical Rashba effect near the Fermi level, and the ratio of ΔE to Δk becomes extremely small.

Among the three monolayers, $TeMoSiN_2$ exhibits the largest intrinsic Rashba constant, with values along the two directions reaching 0.621 eV·Å and 0.582 eV·Å, respectively, as shown in Figure 6a. With the increase in tensile strain, the Rashba constant remains stable along the Γ→K direction, indicating that it has good robustness against uniaxial stretching along this direction.

As shown in Figure 6b, under biaxial strain, a compressive strain of −3% can enhance the Rashba constants of $TeMoSiP_2$ to approximately 0.659 eV·Å and 0.632 eV·Å along the two directions, which are about 1.88 and 1.81 times the intrinsic values, respectively. The Rashba constant of $TeMoGeP_2$ increases continuously with the enhancement of compressive strain; under a compressive strain of −4%, the Rashba constants are increased to 0.782 eV·Å and 0.702 eV·Å. However, both $TeMoGeP_2$ and $TeMoSiN_2$ show a trend of first decreasing and then increasing under tensile strain. For $TeMoSiN_2$, the splitting originally dominated by the Mo-$d_z{}^2$ and N-p_z orbitals gradually transforms into a Mexican hat-type splitting, as shown in Figure 7e,f; the proportion of Mo-$d_z{}^2$ is 41.1% and that of P-p_x is 19.7%.

Most monolayers exhibit more significant anisotropy under uniaxial strain. For example, under a −4% uniaxial compressive strain, the Rashba constants of $TeMoSiP_2$ along the Γ→M and Γ→K directions differ by 0.222 eV·Å. When $TeMoGeP_2$ is subjected to a −2% biaxial compressive strain, this difference is only 0.039 eV·Å, but it reaches 0.182 eV·Å under uniaxial strain.

3.6.2. Electric Field

As illustrated in Figure 8, the Rashba constants of $SeMoGeP_2$ exhibit an almost linear correlation with the applied electric field, where a negative electric field can enhance the Rashba constants to a certain degree. However, such a typical linear relationship is not universal among $XMoYZ_2$ compounds.

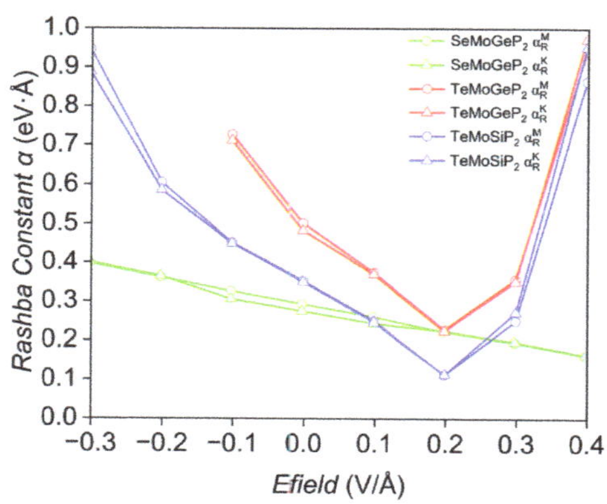

Figure 8. The variation in the Rashba constants with the electric field, the green, red, and blue lines represent $SeMoGeP_2$, $TeMoGeP_2$ and $TeMoSiP_2$, respectively.

Figure 9 depicts the electronic structure of $TeMoGeP_2$ under different electric fields. When the electric field reaches −0.1 V/Å, the Rashba constants increase to 0.727 eV·Å and 0.712 eV·Å, respectively. Beyond this threshold, the P-p_x and P-p_y orbitals which do not exhibit the Rashba effect take the place of the original Mo-$d_z{}^2$ and P-p_z orbital; the proportions of Mo-$d_z{}^2$ and P-p_z change from 31.9% and 15.4% to 0% and 0%, respectively, but P-p_x and P-p_y emerge as 17.3% and 17.4%. A positive electric field induces a non-monotonic variation in the Rashba constants, characterized by an initial decrease followed

 https://doi.org/10.3390/nano16060358

by an increase. As shown in Table 5, in the electric field range of 0 V/Å to 0.2 V/Å, the Δk values along the two directions remain nearly unchanged, while ΔE decreases significantly, resulting in a reduction in the Rashba constants. It can be seen from Figure 9c,d that from 0.2 V/Å to 0.3 V/Å, around the Rashba splitting point dominated by the Mo-$d_z{}^2$ orbital, the proportion of Mo-$d_z{}^2$, P-p_z and Te-p_z change from 46%, 6.9% and 2.6% to 46%, 4.1% and 5.5%; the growth rate of ΔE exceeds that of Δk, leading to an increase in the Rashba constants. Specifically, under an electric field of 0.4 V/Å, the Rashba constant reaches approximately 0.95 eV·Å.

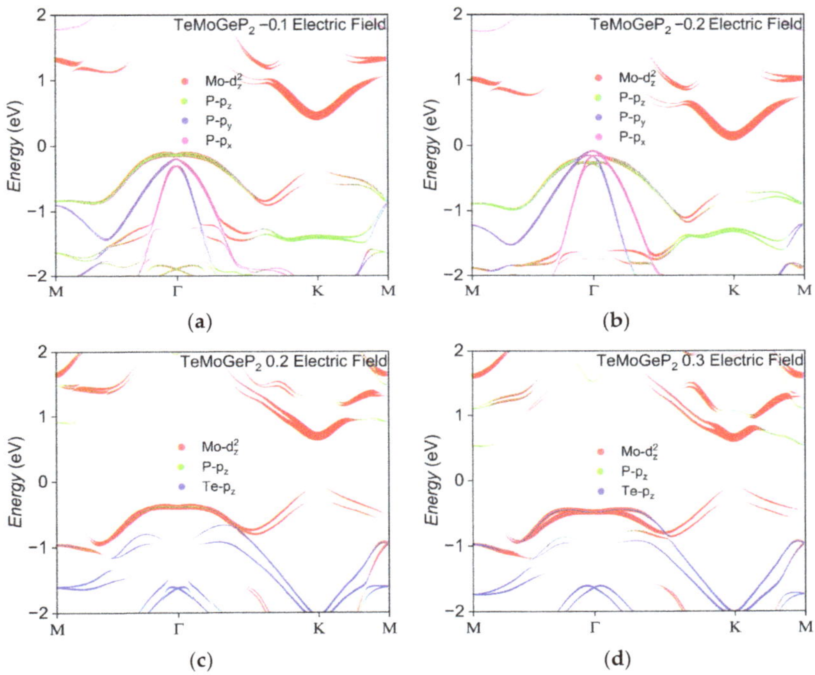

Figure 9. The electronic structure changes of $TeMoGeP_2$ under different electric fields: (**a**) −0.1 V/Å, (**b**) −0.2 V/Å, (**c**) 0.2 V/Å and (**d**) 0.3 V/Å.

Table 5. The variation trends of ΔE and Δk in the electric field of $SeMoGeP_2$, $TeMoGeP_2$ and $TeMoSiP_2$.

	Electric Field (V/Å)	$\Delta E_{\Gamma\to M}$	$\Delta k_{\Gamma\to M}$	$\Delta E_{\Gamma\to K}$	$\Delta k_{\Gamma\to K}$
$SeMoGeP_2$	−0.3	0.01040	0.05215	0.01051	0.05219
	−0.2	0.00946	0.05215	0.00955	0.05219
	−0.1	0.00853	0.05215	0.00861	0.05621
	0	0.00833	0.05215	0.00841	0.05621
	0.1	0.00680	0.05215	0.00686	0.05621
	0.2	0.00583	0.05215	0.00588	0.05219
	0.3	0.00511	0.05215	0.00515	0.05219
	0.4	0.00425	0.05215	0.00428	0.05219
$TeMoGeP_2$	−0.1	0.02853	0.07845	0.02943	0.08271
	0	0.02485	0.09892	0.02568	0.10634
	0.1	0.01980	0.10574	0.02043	0.11028
	0.2	0.01202	0.10574	0.01237	0.11028

https://doi.org/10.3390/nano16060358

Table 5. *Cont.*

		Electric Field (V/Å)			
		$\Delta E_{\Gamma\to M}$	$\Delta k_{\Gamma\to M}$	$\Delta E_{\Gamma\to K}$	$\Delta k_{\Gamma\to K}$
$TeMoGeP_2$	0.3	0.03903	0.21830	0.04359	0.24814
	0.4	0.13178	0.27970	0.16082	0.33085
$TeMoSiP_2$	−0.3	0.04419	0.09346	0.04618	0.10393
	−0.2	0.03358	0.11077	0.03520	0.11992
	−0.1	0.02660	0.11770	0.02787	0.12392
	0	0.02082	0.11770	0.02173	0.12392
	0.1	0.01385	0.11077	0.01434	0.11593
	0.2	0.00509	0.09000	0.00521	0.09194
	0.3	0.02948	0.23540	0.04291	0.31579
	0.4	0.12588	0.29079	0.17120	0.35976

$TeMoSiP_2$ follows a similar variation trend to $TeMoGeP_2$: under an electric field of −0.3 V/Å, Δk decreases while ΔE increases, ultimately leading to Rashba constants along the Γ→K and Γ→M directions of 0.889 eV·Å and 0.946 eV·Å, respectively, 2.53 and 2.70 times the intrinsic values. When the electric field is 0.4 V/Å, the Rashba constants of $TeMoSiP_2$ also reach around 0.9 eV·Å. This represents a substantial enhancement in the Rashba constants of two-dimensional materials.

3.6.3. Charge Doping

Figure 10 illustrates the variations in Rashba constants of the three monolayers under p-doping and n-doping. It is evident that all three monolayers maintain high structural stability during the doping process. For $TeMoGeP_2$, the Rashba constants exhibit a slight increase under n-doping: when the doping level reaches 0.3 e (~0.028 e/Å^2), the Rashba constants increase by 0.017 eV·Å and 0.037 eV·Å, respectively. In contrast, under p-doping (doping level of −0.3 e), the Rashba constants are 0.422 eV·Å and 0.440 eV·Å, corresponding to a decrease of only approximately 0.066 eV·Å. For $TeMoSiN_2$, p-doping can moderately enhance the Rashba constants; specifically, at a doping level of −0.3 e, the Rashba constants along the two directions increase by 0.085 eV·Å and 0.08 eV·Å, respectively. $TeMoSiP_2$ demonstrates high robustness against charge doping.

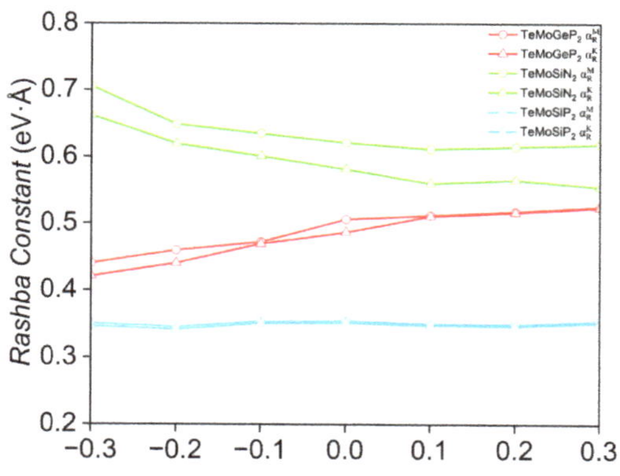

Figure 10. The changes in Rashba constants under charge doping, with the red, green, and blue lines representing $TeMoGeP_2$, $TeMoSiN_2$ and $TeMoSiP_2$, respectively.

 https://doi.org/10.3390/nano16060358

The Fermi level shifts slightly toward the conduction band after n-doping; conversely, p-doping leads to the Fermi level exhibiting a significant shift toward the valence band. The charge doping adjusts the Fermi level by populating or depopulating the electrons near the band edges, but hardly modifies the electronic structures and SOC strength.

We speculate that in $XMoYZ_2$, varying the electron count via charge doping only shifts the Fermi energy significantly, while the Rashba constants remain nearly unchanged and are barely affected by such a moderate charge modulation.

Based on the results, we make the following speculations: the essence of strain is the adjustment of atomic spacing and bond angles; lattice distortion alters the wave function overlap integral of the key orbitals of Rashba splitting; the greater the wave function overlap integral, the higher the probability of electron transitions between different orbitals, the stronger the hybridization, and the higher the coupling efficiency of SOC. For an external electric field, the Coulomb force will change the distribution of the electron cloud, regulating the strength of the vertical intrinsic electric field and the orbital hybridization direction; this belongs to the electron state rearrangement dominated by the Coulomb force. For some $XMoYZ_2$, the modulation has non-monotonicity, related to the direction and intensity of the electric field. The modulation amplitude for $TeMoYZ_2$ is much greater than that for $SMoYZ_2$ or $SeMoYZ_2$. For charge doping, we speculate that only changing the charge density without charge transfer proportion between layers will not change the wave function shape of the orbitals; the overlap integral of Mo-d_{z^2} and Y/Z-p_z orbitals will not change, and the SOC strength remains stable; the Janus structure is not destroyed, the spatial inversion asymmetry remains unchanged, and the intensity of the intrinsic electric field also remains stable.

3.7. Optical Properties

Figure 11 reveals distinct optical absorption responses of $TeMoGeP_2$, $TeMoSiN_2$, and $TeMoSiP_2$, particularly in the visible light and absorption peak regimes. All three monolayers generally exhibit increasingly stronger optical absorption from red light to violet light (about 380 nm−800 nm), with absorption values rapidly rising from 5–10 $\times$ 10^5 to 15–25 $\times$ 10^5 cm^{-1}. The responses of optical absorption to compressive and tensile strains exhibit multiple reversals, providing flexibility for strain modulated optical absorption and indicating optical absorption for visible light applications. Meanwhile, their primary absorption peaks occur between 200 nm and 400 nm, where near-ultraviolet waves are located, reaching maximum intensities of 25–30 $\times$ 10^5 cm^{-1} before gradually declining with increasing energy.

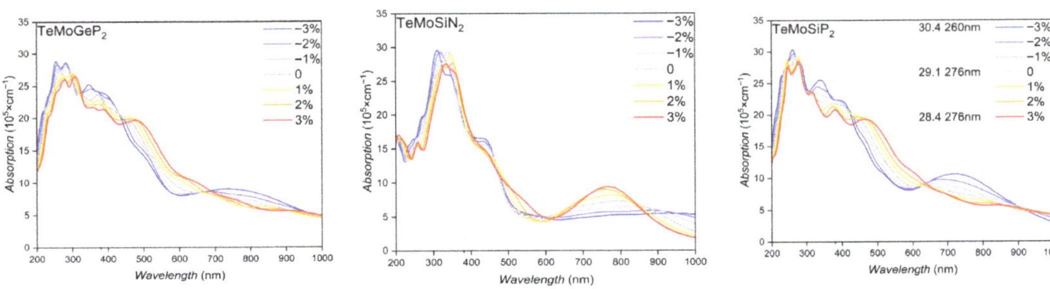

(a) (b) (c)

Figure 11. The optical absorption spectra of $TeMoGeP_2$, $TeMoSiN_2$ and $TeMoSiP_2$ under different levels of biaxial strain. (**a**) $TeMoGeP_2$; (**b**) $TeMoSiN_2$; (**c**) $TeMoSiP_2$.

https://doi.org/10.3390/nano16060358

Compressive strain induces two key effects on the absorption peaks. Firstly, it enhances the peaks' intensity; the blue curves consistently show higher absorption magnitudes than the red curves. Secondly, the absorption peaks shift toward lower wavelength compared to unstrained or tensile strain, so compressive strain may cause blue shifts. This may arise from strain-induced modifications to the electronic structures, where compressive strain may increase the bandgap energy, shifting optical transitions to shorter wavelengths. These findings demonstrate that compressive strain is an effective tool for tuning both the strength and spectral position of absorption peaks in these monolayers.

4. Conclusions

In summary, we performed investigations on the stability, potential, charge transfer, and SOC-considered electronic structure of $XMoYZ_2$ via first-principles calculations. We focused on the modulation of Rashba constants in $XMoYZ_2$ monolayers through strain, external electric fields, and charge doping. The results indicate that $TeMoYZ_2$ exhibits a superior intrinsic Rashba effect, which becomes more prominent under strain modulation. Specifically, a −2% uniaxial strain can increase the Rashba constant of $TeMoGeP_2$ from 0.507 eV·Å to 0.759 eV·Å, and a −4% uniaxial strain can increase the $TeMoSiP_2$ Rashba constant from 0.355 eV·Å to 0.775 eV·Å. Moreover, −0.3 V/Å and 0.4 V/Å external electric fields can raise the Rashba constants of $TeMoSiP_2$ from about 0.35 eV·Å to around/1 eV·Å. $TeMoYZ_2$ demonstrates robustness against charge doping and can slightly enhance the Rashba constants of $TeMoGeP_2$ and $TeMoSiN_2$. Finally, the optical absorption efficiency of $TeMoYZ_2$ was tested and it was found that $TeMoSiN_2$ has strong optical selectivity. In the ultraviolet region, compressive strain can induce a blue shift in the absorption peak and enhance absorption intensity. Through this study, we achieved a more than twofold enhancement in the Rashba constants of several $XMoYZ_2$ materials, which holds substantial promise for applications in electronic devices. The synthesis of $XMoYZ_2$ monolayers is expected via mature chemical vapor deposition, and precise crystallinity is crucial for the predicted topological properties. These findings not only enrich the theoretical basis for regulating the Rashba effect in 2D Janus materials but also provide guidance for the design and development of high-performance spintronic and optoelectronic devices.

Author Contributions: Conceptualization, M.L.; Methodology, M.L.; Software, M.L., P.G., H.S., B.Z. and J.W.; Validation, N.S.; Investigation, X.L., N.S., P.G., H.S., B.Z. and L.L.; Resources, M.L., P.G., H.S., B.Z. and J.W.; Data curation, X.L. and N.S.; Writing—original draft, X.L.; Writing—review & editing, M.L. and J.W.; Visualization, X.L.; Supervision, M.L.; Project administration, M.L.; Funding acquisition, M.L., H.S., B.Z. and J.W. All authors have read and agreed to the published version of the manuscript.

Funding: This research was funded by the Key scientific research projects in higher education institutions in Henan Province grant number 24A140028, by the Funding scheme for young backbone teachers of Henan Province grant number 2023GGJS109 and by the Natural Science Foundation of Henan Province grant number 232300421219 and 252300423539.

Data Availability Statement: Data are contained within the article.

Conflicts of Interest: The authors declare no conflict of interest.

References

1. Lee, S.; Kim, M.; Kwon, Y.K. Unconventional Hidden Rashba Effects in Two-Dimensional InTe. *npj D Mater. Appl.* **2023**, *7*, 43. [CrossRef]
2. Wei, J.; Tian, Q.; Xu, X.; Qin, G.; Zuo, X.; Qin, Z. Two-Dimensional Rashba Semiconductors and Inversion-Asymmetric Topological Insulators in Monolayer Janus $MAA'Z_xZ'_{(4-x)}$ Family. *Appl. Phys. Lett.* **2025**, *126*, 163104. [CrossRef]

https://doi.org/10.3390/nano16060358

3. Sánchez, J.C.R.; Vila, L.; Desfonds, G.; Gambarelli, S.; Attané, J.P.; De Teresa, J.M.; Magén, C.; Fert, A. Spin-to-Charge Conversion Using Rashba Coupling at the Interface between Non-Magnetic Materials. *Nat. Commun.* **2013**, *4*, 2944. [CrossRef] [PubMed]
4. Manchon, A.; Koo, H.C.; Nitta, J.; Frolov, S.M.; Duine, R.A. New Perspectives for Rashba Spin–Orbit Coupling. *Nat. Mater.* **2015**, *14*, 871–882. [CrossRef]
5. Wu, H.; Tian, Q.; Wei, J.; Xing, Z.; Qin, G.; Qin, Z. Rashba Effect Modulation in Two-Dimensional $A_2B_2Te_6$ (A = Sb, Bi; B = Si, Ge) Materials via Charge Transfer. *Nanoscale* **2025**, *17*, 17247–17255. [CrossRef] [PubMed]
6. Bihlmayer, G.; Noël, P.; Vyalikh, D.V.; Chulkov, E.V.; Manchon, A. Rashba-like Physics in Condensed Matter. *Nat. Rev. Phys.* **2022**, *4*, 642–659. [CrossRef]
7. Frank, T.; Faria Junior, P.E.; Zollner, K.; Fabian, J. Emergence of Radial Rashba Spin-Orbit Fields in Twisted van Der Waals Heterostructures. *Phys. Rev. B* **2024**, *109*, L241403. [CrossRef]
8. Kohda, M.; Okayasu, T.; Nitta, J. Spin-Momentum Locked Spin Manipulation in a Two-Dimensional Rashba System. *Sci. Rep.* **2019**, *9*, 1909. [CrossRef]
9. Guo, W.-T.; Huang, Z.; Zhang, J.-M. Rashba Spin-Splitting Driven Inverse Spin Hall Effect in $MnBi_2Te_4$. *Commun. Phys.* **2025**, *8*, 1. [CrossRef]
10. Szczęśniak, D.; Kais, S. Gap States and Valley-Spin Filtering in Transition Metal Dichalcogenide Monolayers. *Phys. Rev. B* **2020**, *101*, 115423. [CrossRef]
11. Koo, H.C.; Kim, S.B.; Kim, H.; Park, T.; Choi, J.W.; Kim, K.; Go, G.; Oh, J.H.; Lee, D.; Park, E.; et al. Rashba Effect in Functional Spintronic Devices. *Adv. Mater.* **2020**, *32*, 2002117. [CrossRef]
12. Cheng, C.; Sun, J.-T.; Chen, X.-R.; Fu, H.-X.; Meng, S. Nonlinear Rashba Spin Splitting in Transition Metal Dichalcogenide Monolayers. *Nanoscale* **2016**, *8*, 17854–17860. [CrossRef]
13. Xiang, L.; Ke, Y.; Zhang, Q. Tunable giant Rashba-type spin splitting in $PtSe_2/MoSe_2$ heterostructure. *Appl. Phys. Lett.* **2019**, *115*, 203501. [CrossRef]
14. Zhang, L.; Gu, Y.; Du, A. Two-Dimensional Janus Antimony Selenium Telluride with Large Rashba Spin Splitting and High Electron Mobility. *ACS Omega* **2021**, *6*, 31919–31925. [CrossRef] [PubMed]
15. Chen, J.; Wu, K.; Ma, H.; Hu, W.; Yang, J. Tunable Rashba Spin Splitting in Janus Transition-Metal Dichalcogenide Monolayers via Charge Doping. *RSC Adv.* **2020**, *10*, 6388–6394. [CrossRef] [PubMed]
16. Zubair, M.; Vasilopoulos, P.; Tahir, M. Influence of Interface Induced Valley-Zeeman and Spin-Orbit Couplings on Transport in Heterostructures of Graphene on WSe_2. *Phys. Rev. B* **2020**, *101*, 165436. [CrossRef]
17. Rani, R.; Sharma, M.; Sharma, R. Giant Quantum Capacitance and Rashba Splitting in Tellurene Bilayer Derivatives. *Physica E* **2023**, *152*, 115757. [CrossRef]
18. Gupta, S.; Yakobson, B.I. What Dictates Rashba Splitting in 2D van Der Waals Heterobilayers. *J. Am. Chem. Soc.* **2021**, *143*, 3503–3508. [CrossRef]
19. Bordoloi, A.; Garcia-Castro, A.C.; Romestan, Z.; Romero, A.H.; Singh, S. Promises and Technological Prospects of Two-Dimensional Rashba Materials. *Appl. Phys.* **2024**, *135*, 220901. [CrossRef]
20. Affandi, Y.; Ulil Absor, M.A. Electric Field-Induced Anisotropic Rashba Splitting in Two-Dimensional Tungsten Dichalcogenides WX_2 (X: S, Se, Te): A First-Principles Study. *Physica E* **2019**, *114*, 113611. [CrossRef]
21. Nayak, C.; Masanta, S.; Ghosh, S.; Moulick, S.; Pal, A.N.; Bose, I.; Singha, A. Valley Polarization and Photocurrent Generation in Transition Metal Dichalcogenide Alloy $MoS_{2x}Se_{2(1-x)}$. *Phys. Rev. B* **2024**, *109*, 115304. [CrossRef]
22. Zamanian, E.; Babaee Touski, S. Electronic, Rashba and Photocatalytic Properties of Janus $XMoYZ_2$ (X = S, Se, Te; Y = Si, Ge and Z = N, P) Monolayers. *Physica E* **2024**, *163*, 116012. [CrossRef]
23. Yu, S.-B.; Zhou, M.; Zhang, D.; Chang, K. Spin Hall Effect in the Monolayer Janus Compound MoSSe Enhanced by Rashba Spin-Orbit Coupling. *Phys. Rev. B* **2021**, *104*, 075435. [CrossRef]
24. Lee, S.; Kwon, Y.-K. Unveiling Giant Hidden Rashba Effects in Two-Dimensional Si_2Bi_2. *npj 2D Mater. Appl.* **2020**, *4*, 45. [CrossRef]
25. Freitas, R.R.Q.; De Brito Mota, F.; Rivelino, R.; De Castilho, C.M.C.; Kakanakova-Georgieva, A.; Gueorguiev, G.K. Spin-Orbit-Induced Gap Modification in Buckled Honeycomb XBi and XBi_3 (X = B, Al, Ga, and In) Sheets. *J. Phys. Condens. Matter* **2015**, *27*, 485306. [CrossRef]
26. Zhang, S.-H.; Liu, B.-G. Anisotropic Rashba Effect and Charge and Spin Currents in Monolayer BiTeI by Controlling Symmetry. *Phys. Rev. B* **2019**, *100*, 165429. [CrossRef]
27. Sibatov, R.T.; Meftakhutdinov, R.M.; Kochaev, A.I. Asymmetric $XMoSiN_2$ (X = S, Se, Te) Monolayers as Novel Promising 2D Materials for Nanoelectronics and Photovoltaics. *Appl. Surf. Sci.* **2022**, *585*, 152465. [CrossRef]
28. Farooq, M.U.; Xian, L.; Huang, L. Spin Hall Effect in Two-Dimensional InSe: Interplay between Rashba and Dresselhaus Spin-Orbit Couplings. *Phys. Rev. B* **2022**, *105*, 245405. [CrossRef]
29. Rezavand, A.; Ghobadi, N. Stacking-Dependent Rashba Spin-Splitting in Janus Bilayer Transition Metal Dichalcogenides: The Role of in-Plane Strain and out-of-Plane Electric Field. *Physica E* **2021**, *132*, 114768. [CrossRef]

 https://doi.org/10.3390/nano16060358

30. Andrews, N.L.P.; Fan, J.Z.; Forward, R.L.; Chen, M.C.; Loock, H.-P. Ideal strength and strain engineering of the Rashba effect in two-dimensional BiTeBr. *Phys. Chem. Chem. Phys.* **2017**, *19*, 73–81. [CrossRef] [PubMed]
31. Hu, T.; Jia, F.; Zhao, G.; Wu, J.; Stroppa, A.; Ren, W. Intrinsic and Anisotropic Rashba Spin Splitting in Janus Transition-Metal Dichalcogenide Monolayers. *Phys. Rev. B* **2018**, *97*, 235404. [CrossRef]
32. Singh, S.; Romero, A.H. Giant Tunable Rashba Spin Splitting in a Two-Dimensional BiSb Monolayer and in BiSb/AlN Heterostructures. *Phys. Rev. B* **2017**, *95*, 165444. [CrossRef]
33. Su, J.; Wang, X.; Shao, C.; Guo, Y.; Xian, L. Strain-Controlled Rashba Spin-Orbit Coupling Effect in SnS and SnSe Monolayers. *Mater. Des.* **2021**, *209*, 110005. [CrossRef]
34. Hong, Y.L.; Liu, Z.; Wang, L.; Zhou, T.; Ma, W.; Xu, C.; Feng, S.; Chen, L.; Chen, M.L.; Sun, D.M.; et al. Chemical vapor deposition of layered two-dimensional $MoSi_2N_4$ materials. *Science* **2020**, *369*, 670–674. [CrossRef] [PubMed]
35. Dai, Z.; Liu, L.; Zhang, Z. Strain Engineering of 2D Materials: Issues and Opportunities at the Interface. *Adv. Mater.* **2019**, *31*, 1805417. [CrossRef] [PubMed]
36. Kresse, G.; Furthmüller, J. Efficient Iterative Schemes for Ab Initio Total-Energy Calculations Using a Plane-Wave Basis Set. *Phys. Rev. B* **1996**, *54*, 11169–11186. [CrossRef]
37. Perdew, J.P.; Burke, K.; Ernzerhof, M. Generalized Gradient Approximation Made Simple. *Phys. Rev. Lett.* **1996**, *77*, 3865–3868. [CrossRef]
38. Blöchl, P.E. Projector Augmented-Wave Method. *Phys. Rev. B* **1994**, *50*, 17953–17979. [CrossRef]
39. Bader, R.F.W.; Henneker, W.H.; Cade, P.E. Molecular Charge Distributions and Chemical Binding. *Chem. Phys.* **1967**, *46*, 3341–3363. [CrossRef]
40. Togo, A.; Chaput, L.; Tadano, T.; Tanaka, I. Implementation Strategies in Phonopy and Phono3py. *J. Phys. Condens. Matter* **2023**, *35*, 353001. [CrossRef]
41. Baroni, S.; De Gironcoli, S.; Dal Corso, A.; Giannozzi, P. Phonons and Related Crystal Properties from Density-Functional Perturbation Theory. *Rev. Mod. Phys.* **2001**, *73*, 515–562. [CrossRef]
42. Togo, A.; Tanaka, I. First Principles Phonon Calculations in Materials Science. *Scr. Mater.* **2015**, *108*, 1–5. [CrossRef]
43. Pallikara, I.; Kayastha, P.; Skelton, J.M.; Whalley, L.D. The Physical Significance of Imaginary Phonon Modes in Crystals. *Electron. Struct.* **2022**, *4*, 033002. [CrossRef]

 https://doi.org/10.3390/nano16060358

nanomaterials

Article

In Silico Models for Predicting Adsorption of Organic Pollutants on Atmospheric Nanoplastics by Combining Grand Canonical Monte Carlo/Density Functional Theory and Quantitative Structure Activity Relationship Approach

Ya Wang [1], Honghong Yi [1], Chao Li [2,3,*], Xiaolong Tang [1], Peng Zhao [1] and Zhongfang Chen [4,*]

1 School of Energy and Environmental Engineering, University of Science and Technology Beijing, Beijing 100083, China; wyanne@ustb.edu.cn (Y.W.)
2 Engineering Laboratory for Water Pollution Control and Resources Recovery, School of Environment, Northeast Normal University, Changchun 130117, China
3 State Environmental Protection Key Laboratory of Wetland Ecology and Vegetation Restoration, School of Environment, Northeast Normal University, Changchun 130117, China
4 Department of Chemistry, University of Puerto Rico, San Juan, PR 00931, USA
* Correspondence: lic932@nenu.edu.cn (C.L.); zhongfang.chen1@upr.edu (Z.C.)

Abstract

Estimating the adsorption data and understanding the adsorption behavior and mechanism of organic pollutants on nanoplastics are crucial for assessing their ecological risks. Herein, in silico techniques, i.e., grand canonical Monte Carlo simulations, density functional theory computations, and quantitative structure activity relationship (QSAR) modeling, were integrated to examine the adsorption of 39 representative aliphatic and aromatic compounds and nine emerging pollutants (brominated flame retardants and phosphorus flame retardants) onto 12 different nanoplastics under atmospheric conditions. Three QSAR models were constructed to predict the adsorption equilibrium constant (log$K$) for polyethylene, polyoxymethylene, and polyvinyl alcohol nanoplastics individually, along with 12 QSAR models for separately estimating adsorption capacities (C_m) on different nanoplastics. Furthermore, a novel multi-dimensional prediction model was developed, enabling simultaneous, high-throughput prediction of adsorption capacities across multiple nanoplastics and pollutants with a single input. These results revealed that van der Waals and electrostatic interactions serve as the primary driving forces for the adsorption. The novel multi-dimensional prediction model facilitates rapid and comprehensive assessment of pollutant–nanoplastic interactions with one-click, and paves the way for improved risk evaluations and advancing predictive environmental research.

Keywords: nanoplastics; adsorption equilibrium constant; adsorption capacity; grand canonical Monte Carlo (GCMC) simulation; density functional theory (DFT) calculations; predictive model

1. Introduction

The synthesis of plastics dates back to the early 20th century, and their production and use have grown exponentially since the 1950s [1]. It is estimated that the global plastic production exceeded 400 million metric tons (Mt) in 2022, and will more than double by 2050 compare d with 2020 levels [2,3]. Alarmingly, only ~9% of plastics are recycled and ~12% are incinerated, leaving ~79% deposited of in landfills or released into the environment [1,4]. Through natural weathering in the environment, larger plastic debris progressively fragments into microplastics (1 μm to 5 mm) and ultimately into nanoplastics (NPs, <1 μm) [5], which have been widely detected across various environmental media, including water, air, soil, and biosphere [6–8]. Owing to their hydrophobic nature and high specific surface area, NPs tend to adsorb organic pollutants, and can act as vectors that facilitate pollutant transport in the environment [9–11], thereby influencing pollutant fate and potentially generating ecological risks. Therefore, it is critical to understand the adsorption mechanisms of organic pollutants on NPs, and evaluate their environmental behavior and associated ecological risks.

Despite its importance, research on the adsorption of organic pollutants onto NPs is still in its infancy, with limited experimental data available. Existing experimental studies have investigated the adsorption of various organic compounds onto diverse microplastics/NPs in an aqueous environment, uncovering heterogeneous and complex adsorption mechanisms. A single microplastic or NP can exhibit varying adsorption capabilities for different organic pollutants. For instance, the adsorption capacities of polyethylene microplastic increase sequentially for carbamazepine, 17α-ethinyl estradiol, triclosan, and 4-methylbenzylidene camphor [12]. Similarly, different microplastics/NPs show different adsorption capabilities for the same pollutant. For example, polystyrene nanoplastic exhibits stronger adsorption for polychlorinated biphenyl compared to polyethylene microplastic [13]. Additionally, surface functionalization of plastics, such as functional groups or attached biofilms, can affect the adsorption behavior [14,15]. Moreover, environmental factors, such as temperature [16], pH [17], and dissolved organic matter [18] in the water environment also play critical roles.

Typically, multiple interactions including π-π interactions, hydrophobic forces, dispersion, electrostatic interactions, and hydrogen bonding coexist during the adsorption of organic pollutants onto microplastics/NPs, and these interactions depend on the structures and properties of both microplastics/NPs and organic pollutants [19–21]. Significant insights have been gained from studies in aqueous environments [22,23], while research on how the structures and properties of NPs and organic pollutants influence these interactions in the atmosphere remains limited. Therefore, it is crucial to elucidate the adsorption mechanism of diverse organic pollutants on NPs in the atmospheric conditions.

Compared with experimental methods, computational approaches offer significant advantages for elucidating adsorption mechanisms at the atomic level. These techniques provide detailed insights into the interactions between NPs and pollutants, enabling systematic interrogation of adsorption processes that are often challenging to isolate or directly observe experimentally. Therefore, computational investigations of pollutant adsorption on NPs have gained increasing popularity and represent a growing frontier in environmental research [24].

Various computational methods, such as grand canonical Monte Carlo (GCMC), quantum chemistry, and molecular dynamics, have been applied to explore the adsorption of organic pollutants on NPs, including polyethylene, polypropylene, and polyvinyl chloride [25,26]. Each method offers distinct characteristics and advantages. GCMC simulations [27,28] can be employed to obtain the adsorption isotherms and predict adsorption capacities based on the probabilistic statistical principles; however, they cannot provide direct insights into the adsorption dynamics. In contrast, molecular dynamics simulations enable the modeling of dynamic adsorption processes. However, both GCMC and molecular dynamics cannot provide accurate electronic properties, while quantum chemical approaches, particularly density-functional theory (DFT) [29,30], can provide us accurate adsorption energies along detailed electronic properties. Nevertheless, DFT calculations entail higher computational costs than either GCMC or molecular dynamics simulations.

https://doi.org/10.3390/nano16030178

Given the vast number of organic pollutants and diverse NPs in the environment, it is not practical, if not impossible, to simulate the adsorption process individually for each pollutant–NP pair. Hence, it is urgent to leverage simulation methods and develop predictive models to efficiently and accurately estimate adsorption data, thereby deepening our understanding of pollutant–NP interactions in the atmosphere. Quantitative structure activity relationship (QSAR) techniques have been successfully utilized to establish predictive models for adsorption of organic pollutants on microplastics in aquatic environments [31–35]. However, these models are largely restricted to aqueous systems, and equivalent models tailored to gas-phase atmospheric conditions remain scarce. Therefore, there is an urgent need to develop novel predictive models to estimate the adsorption of organic pollutants towards NPs in the atmosphere.

In this work, 39 representative aliphatic and aromatic compounds, along with nine emerging pollutants, including brominated flame retardants and phosphorus flame retardants, were selected as adsorbates, while 12 different nanoplastics were chosen as adsorbents. Using GCMC simulations, we estimated the adsorption capacities (C_m) of these 48 organic pollutants on the 12 NPs under atmospheric conditions. Subsequently, DFT calculations were carried out to obtain the logarithm of the adsorption equilibrium coefficients (log$K$) of organic pollutants on three NPs, namely, polyethylene (PE), polyoxymethylene (POM), and polyvinyl alcohol (PVA), since they exhibited the highest adsorption capacities among these 12 NPs. We also developed three QSAR models for predicting the log$K$ values of organic pollutants on PE, POM and PVA individually, as well as 12 QSAR models for predicting the C_m values for each NP. Based on these 12 QSAR models, we established a novel multi-dimensional QSAR model incorporating parameters describing the structures of both organic pollutants and NPs, enabling simultaneous prediction of adsorption capacities across diverse organic pollutants and NPs in the atmosphere. These QSAR models not only advance mechanistic understanding of adsorption on NPs in atmospheric environments, but also provide efficient tools for generating adsorption data that are critical for assessing the ecological risks associated with organic pollutants and NPs.

2. Computational Details

2.1. Adsorbent and Adsorbate Models

Twelve different nanoplastics (Figure 1) were constructed as adsorbent models in a periodic simulation cell of 25 Å × 25 Å × 50 Å. Each nanoplastic structure was designed with the corresponding cell length of 50 Å along the Z-direction so as to provide sufficient gas-phase space for adsorption sampling and to reduce interactions between periodic images along the surface-normal direction. As exhibited in Figure 1, among these twelve nanoplastics, PC, PS and PET contain phenyl rings, while the remaining nine nanoplastics do not. All nanoplastic structures were fully geometry-relaxed prior to GCMC simulations; the non-linear conformation of PVA originates from the stabilized hydrogen-bonding network of hydroxyl groups, rather than from particle deformation or clustering".

The adsorbates comprised 48 different organic compounds, including 39 representative aliphatic and aromatic compounds with different functional groups, i.e., -OH, $-NH_2$, $-NO_2$, $-CH_3$, -CN, -CHO, $-CH_2OH$, $-CH_2CH_3$, $-C(O)CH_3$, $-CH_2CH_2OH$, $-OC(O)CH_3$, $-C(O)OCH_3$, $-CH_2CH_2CH_3$, $-C(O)OCH_2CH_3$, -F and $-C_6H_5$, and nine emerging pollutants (including brominated flame retardants and phosphorus flame retardants), as listed in Table S1 of the Supplementary Material.

https://doi.org/10.3390/nano16030178

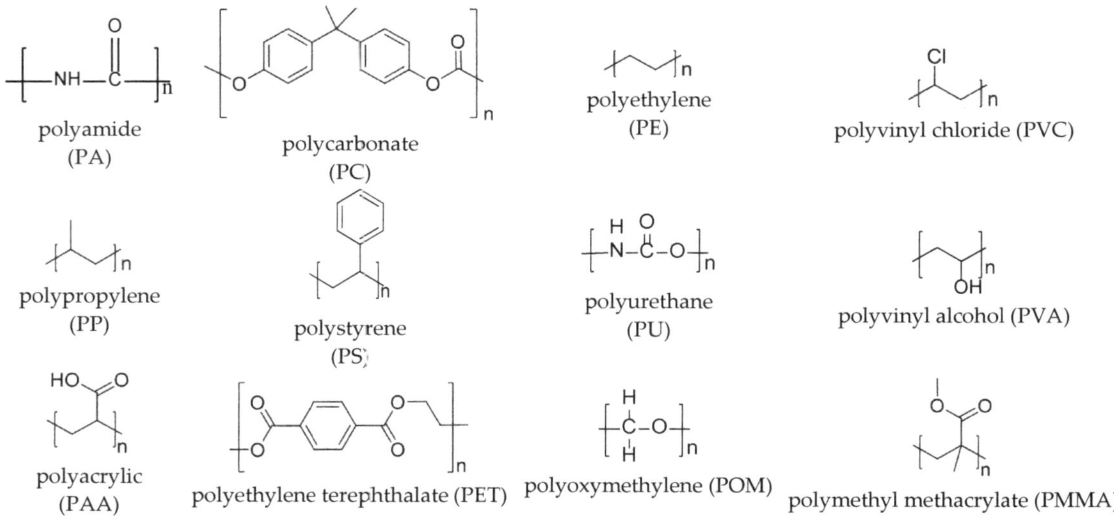

Figure 1. Twelve model nanoplastics.

2.2. Grand Canonical Monte Carlo Simulation

All the GCMC simulations were conducted under the grand canonical ensemble using the Sorption program [36]. The Metropolis algorithm [37] was used to ensure sufficient sampling. The COMPASS II forcefield [38], which extends the coverage for the COMPASS force field [39–41] to polymer and drug-like molecules, was utilized for all the adsorption isotherm simulations. Fugacity values ranged from 101.325 kPa to 1013.25 kPa, and the temperature was 298 K. The upper fugacity values were used to construct complete adsorption isotherms and identify saturation behavior, rather than to represent typical ambient atmospheric pressure. All pollutants and nanoplastics were evaluated under the same gas-phase conditions (298 K; no solvent; pH not applicable), and the 101.325 kPa fugacity is used for environmental discussion unless mentioned otherwise. The electrostatic interactions were calculated using the Ewald & Group based method with a cutoff distance of 12.5 Å [42], while the van der Waals interactions were obtained using the atom-based method with the same cutoff distance of 12.5 Å and a spline width of 1 Å [38,43–46]. All Lennard–Jones parameters and fixed partial charges were assigned from the COMPASS force field without reparameterization.

2.3. Density Functional Theory Computation

Among these 12 nanoplastics, those with adsorption capacities (C_m) in the top 30%, as determined by GCMC simulations, were selected for further estimating their adsorption equilibrium coefficient (K) values using DFT calculations. All DFT calculations were carried out with the DMol^3 program [47,48]. The default temperature was set to 0 K, and no solvent effects was considered for these DFT computations. The Perdew–Burke–Ernzerhof (PBE) functional within the generalized gradient approximation (GGA) was employed [49], along with the double-numerical basis with polarization functions (DNP) [50,51], which is comparable to Gaussian 6-31G(d, p) [52]. Additionally, to account for long-range interactions, the PBE + D2 method with the Grimme van der Waals (vdW) correction [53] was applied.

2.4. Adsorption Data and Molecular Structure Descriptors

To examine the adsorption behavior, we calculated both C_m and K values. C_m represents the mass of organic compounds adsorbed per unit mass of the adsorbent, and

https://doi.org/10.3390/nano16030178

can be obtained from the GCMC simulations. In contrast, the K values, representing the adsorption equilibrium coefficient, can be calculated as follows [54]:

$$K = e^{-\frac{\Delta G}{RT}} \quad (1)$$

where T is the temperature, set at 298.15 K; R is the universal gas constant (8.314 $J \cdot mol^{-1} \cdot K^{-1}$). ΔG represents the change in Gibbs free energy, and can be estimated with the following equation [55]:

$$\Delta G = \Delta E + \Delta ZPE - T\Delta S \quad (2)$$

where ΔE is the changes in total energy, ΔZPE is the changes in zero-point energy, and ΔS is the changes in entropy. ΔE, ΔZPE, and ΔS can be obtained from the DFT computations.

The molecular structures of the 12 nanoplastic monomers and 48 organic pollutants were fully optimized using DFT. Based on these 48 optimized pollutant structures, 2325 molecular structural descriptors and fingerprints were generated using PaDEL software (http://padel.nus.edu.sg/software/padeldescriptor, accessed on 17 May 2024) [56]. Descriptors and fingerprints with a zero value were removed, resulting in 643 relevant parameters for developing QSAR models. Similarly, we calculated the descriptors and fingerprints for the 12 nanoplastics using their optimized structures. These descriptors and fingerprints encode the structural characteristics of these 12 nanoplastics and were therefore used in establishing the QSAR models.

2.5. Establishment and Evaluation for QSAR Models

For the development of QSAR models to predict log$K$ values, these 48 organic compounds were randomly divided into a training set of 40 compounds and a validation set of eight compounds. In contrast, for the QSAR models predicting C_m values, all the 48 organic compounds were used to construct the models without external validation. Multiple linear regression (MLR) analysis was conducted to construct QSAR models. We further evaluated the goodness-of-fit, robustness, and predictive capability of the established QSAR models using standard statistic metrics, including the adjusted determination coefficient (R^2_{adj}), root mean square error for the training set ($RMSE_t$), and, for the validation set, $RMSE_v$, the external determination coefficient (R^2_{ext}), and the external explained variance (Q^2_{ext}).

Additionally, we used Williams plots, which are based on standardized residuals (δ^*) and leverage values (h_i) [57], to characterize the application domain (AD) of the developed prediction models.

3. Results and Discussion

3.1. C_m and logK Values for Organic Compounds on Nanoplastics

As shown in Table S1 and Figure 2, the C_m values for 48 organic compounds towards 12 nanoplastics range from 1.58 to 150.80 g/g. Among these nanoplastics, polyethylene (PE), polyoxymethylene (POM), and polyvinyl alcohol (PVA) (Figure 3), exhibit C_m values within the top 30%, demonstrating their superior adsorption capability compared to the other nanoplastics examined. The C_m values (g/g) for PE are in the range of 5.60~150.80 g/g, while those for POM range from 15.90 to 55.66 g/g, and for PVA, from 4.99 to 43.75 g/g. To better understand the adsorptions on these three MPs, Figure 3 presents the isodensity surfaces for benzene adsorption, colored by potential energies, revealing distinct favorable binding sites for benzene on PE, POM and PVA within their respective polymer frameworks.

https://doi.org/10.3390/nano16030178

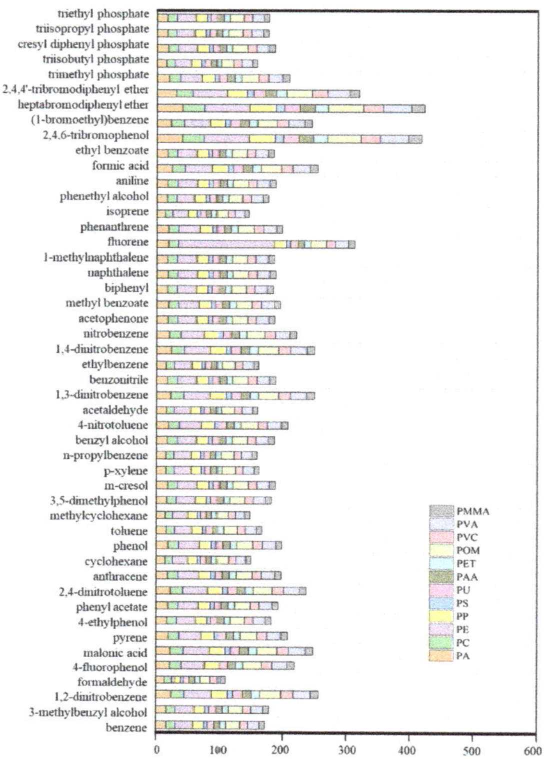

Figure 2. C_m values (g/g) for 48 organic compounds on 12 nanoplastics from GCMC simulations.

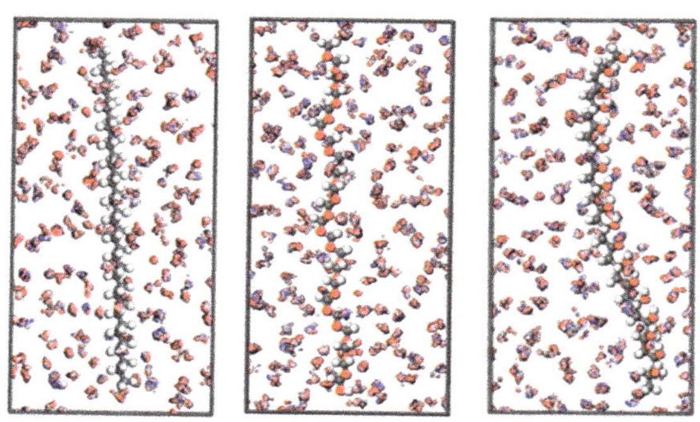

(a) PE (b) POM (c) PVA

Figure 3. Representative isodensity surfaces colored with the potential energies for benzene adsorption on (**a**) PE, (**b**) POM, and (**c**) PVA obtained from GCMC simulations (The dark red and blue areas denote the highest and lowest energy, respectively).

Furthermore, we computed the adsorption of 48 organic compounds on PE, POM, and PVA nanoplastics using DFT, and obtained the corresponding log*K* values. In our previous studies [58,59], we have validated the reliability of adsorption data from DFT computations. Besides, we compared the gas-phase log*K* values for five organic compounds (i.e., benzene, naphthalene, anthracene, nitrobenzene and phenol) on PE MPs from our DFT computations with those from molecular dynamics (MD) simulations [60], for which the methodology

 https://doi.org/10.3390/nano16030178

had been benchmarked against experimental aqueous log$K$ data. These log$K$ values from DFT computations agreed well with those from the verified MD simulations (Table S2). Thus, the atmospheric log$K$ values obtained in this study (Table S1) can be considered as robust and reliable. The optimized adsorption equilibrium configurations for the 48 organic compounds on PE, POM, and PVA nanoplastics are presented in Table S3. The distances between the molecular mass centers and the nanoplastics range from 2.87 to 5.91 Å.

3.2. Sorbent-Specific QSAR Models for Predicting logK Values on PE, POM, and PVA Nanoplastics

The optimal prediction models for estimating the log$K$ values on PE, POM, and PVA nanoplastics were as follows:

for PE nanoplastics,

$$\log K = -10.626 + 2.445 \times ATSC0p + 4.041 \times SpMax1\_Bhm - 1.261 \times LipoaffinityIndex$$

$$n_t = 40, R^2_{\text{adj}} = 0.95, RMSE_{\text{t}} = 2.32, F = 257.17, p < 0.001, \quad (3)$$

$$n_{\text{v}} = 8, R^2_{\text{ext}} = 0.92, Q^2_{\text{ext}} = 0.91, RMSE_{\text{V}} = 3.09$$

for POM nanoplastics,

$$\log K = -6.299 + 2.17 \times ATSC0p + 3.453 \times SpMax1\_Bhm - 1.266 \times LipoaffinityIndex$$

$$n_t = 40, R^2_{\text{adj}} = 0.93, RMSE_{\text{t}} = 2.53, F = 163.77, p < 0.001, \quad (4)$$

$$n_{\text{v}} = 8, R^2_{\text{ext}} = 0.97, Q^2_{\text{ext}} = 0.96, RMSE_{\text{V}} = 1.74$$

and for PVA nanoplastics,

$$\log K = -9.963 + 2.879 \times ATSC0p + 4.168 \times SpMax1\_Bhm - 2.246 \times LipoaffinityIndex$$

$$n_t = 40, R^2_{\text{adj}} = 0.91, RMSE_{\text{t}} = 3.57, F = 139.02, p < 0.001, \quad (5)$$

$$n_{\text{v}} = 8, R^2_{\text{ext}} = 0.94, Q^2_{\text{ext}} = 0.91, RMSE_{\text{V}} = 3.23$$

Here, n_t and n_{v} represent the number of organic compounds in the training set and validation set, respectively. To evaluate the goodness-of-fit, robustness and predictive capability, we calculated a few parameters, including R^2_{adj}, R^2_{ext}, Q^2_{ext}, $RMSE_{\text{t}}$, and $RMSE_{\text{V}}$. According to the established criteria ($R^2 > 0.60$ and $Q^2 > 0.50$) [61], these three QSAR models perform well. Additionally, the variable inflation factors (*VIF*) values for the predictive variables (Table S4) used in these three QSAR models are less than 10, indicating that no serious multi-collinearity exists among these variables [62]. Figure 4 compares the QSAR-predicted log$K$ values with the DFT-calculated log$K$ values for PE, POM, and PVA, demonstrating the predictive accuracy of the developed models. The close agreement indicates that these QSAR models provide a reliable and efficient alternative to additional simulations or experiments for rapidly estimating logK values of organic pollutants on PE, POM, and PVA nanoplastics.

Then, we characterized the application domains (ADs) of the developed QSAR models on the basis of δ^* and h_{i} values (Figure S1). All standardized residuals for the organic

https://doi.org/10.3390/nano16030178

compounds in both the training set and the validation set are in the range of −3 to 3, which implies that no outlier exists. Note that the leverage value for one compound in the validation, namely, cresyl diphenyl phosphate, exceeds the warning leverage value of 0.3, which suggests that the structure of cresyl diphenyl phosphate significantly differs from those of the organic compounds in the training set. Nevertheless, the predicted log*K* values for cresyl diphenyl phosphate on nanoplastics from the QSAR models align well with the DFT values, indicating that the established QSAR models possess reliable extrapolation capabilities. Consequently, these three QSAR models can reliably estimate adsorption on PE, POM, and PVA nanoplastics for a wide range of organic compounds, including alkanes, alkenes, aromatic hydrocarbons, alcohols, ketones, aldehydes, carboxylic acids, phenols, anilines, nitriles, nitrobenzenes, esters, polycyclic aromatic hydrocarbons, biphenyls, halogenated benzenes, polybrominated diphenyl ethers, and organophosphates.

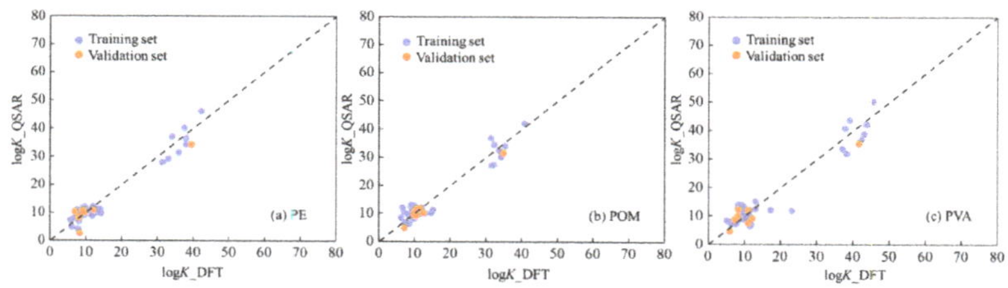

Figure 4. Predicted log*K* values with QSAR models (log*K*_QSAR) versus those from DFT calculations (log*K*_DFT).

3.3. Multi-Dimensional QSAR Models for Predicting C_m Values on Nanoplastics

Note that each of the aforementioned QSAR models for predicting log*K* values is specific to a particular nanoplastic. To predict the log*K* values on different nanoplastics, separate models must be established for each material. This limitation underscores the necessity of developing a multi-dimensional predictive model capable of efficiently generating adsorption data across diverse nanoplastics. Considering computational costs and data volume, we focused on predicting C_m values as an example to explore the development of a multi-dimensional prediction model.

Firstly, we developed 12 QSAR models using the C_m values and structural descriptors of 48 organic compounds. Afterward, a multi-dimensional prediction model was constructed by integrating these 12 QSAR models with the structural descriptors of the 12 nanoplastics.

3.3.1. Twelve Sorbent-Specific QSAR Models

QSAR models for predicting the C_m values on 12 nanoplastics are shown below.

For PA nanoplastics,

$$C_m = 8.108 - 0.017 \times AATS5m + 0.395 \times MATS8c + 0.0446 \times GATS8c - 4.639 \times VE1\_DzZ + 1.556 \times AMW \quad (6)$$
$$n = 48,\ R^2_{adj} = 0.94,\ RMSE = 1.27,\ F = 148.69,\ p < 0.001$$

For PC nanoplastics,

$$C_m = 6.706 - 0.014 \times AATS5m + 0.043 \times MATS8c + 0.456 \times GATS8c - 3.686 \times VE1\_DzZ + 1.289 \times AMW \quad (7)$$
$$n = 48,\ R^2_{adj} = 0.94,\ RMSE = 1.05,\ F = 149.46,\ p < 0.001$$

For PE nanoplastics,

$$C_m = 12.592 - 0.018 \times AATS5m + 75.837 \times MATS8c - 25.79 \times GATS8c - 46.757 \times VE1\_DzZ + 2.86 \times AMW \quad (8)$$
$$n = 48,\ R^2_{adj} = 0.79,\ RMSE = 8.58,\ F = 36.01,\ p < 0.001$$

https://doi.org/10.3390/nano16030178

For PP nanoplastics,

$$C_m = 8.226 - 0.017 \times AATS5m + 0.511 \times MATS8c + 0.559 \times GATS8c - 7.509 \times VE1\_DzZ + 1.551 \times AMW \quad n = 48, R^2_{adj} = 0.88, RMSE = 1.86, F = 71.33, p < 0.001 \tag{9}$$

For PS nanoplastics,

$$C_m = 2.647 - 0.006 \times AATS5m + 0.145 \times MATS8c + 0.196 \times GATS8c - 1.06 \times VE1\_DzZ + 0.52 \times AMW \quad n = 48, R^2_{adj} = 0.77, RMSE = 0.89, F = 31.95, p < 0.001 \tag{10}$$

For PU nanoplastics,

$$C_m = 4.904 - 0.010 \times AATS5m + 0.09 \times MATS8c + 0.216 \times GATS8c - 2.58 \times VE1\_DzZ + 0.922 \times AMW \quad n = 48, R^2_{adj} = 0.94, RMSE = 0.56, F = 150.16, p < 0.001 \tag{11}$$

For PAA nanoplastics,

$$C_m = 4.718 - 0.01 \times AATS5m + 0.082 \times MATS8c + 0.183 \times GATS8c - 2.394 \times VE1\_DzZ + 0.9 \times AMW \quad n = 48, R^2_{adj} = 0.95, RMSE = 0.68, F = 172.98, p < 0.001 \tag{12}$$

For PET nanoplastics,

$$C_m = 4.317 - 0.011 \times AATS5m + 0.143 \times MATS8c + 0.213 \times GATS8c - 2.491 \times VE1\_DzZ + 0.849 \times AMW \quad n = 48, R^2_{adj} = 0.94, RMSE = 0.64, F = 153.35, p < 0.001 \tag{13}$$

For POM nanoplastics,

$$C_m = 10.747 - 0.019 \times AATS5m + 0.3 \times MATS8c + 0.543 \times GATS8c - 7.442 \times VE1\_DzZ + 1.99 \times AMW \quad n = 48, R^2_{adj} = 0.93, RMSE = 1.88, F = 122.73, p < 0.001 \tag{14}$$

For PVC nanoplastics,

$$C_m = 6.189 - 0.013 \times AATS5m + 0.093 \times MATS8c + 0.384 \times GATS8c - 3.872 \times VE1\_DzZ + 1.174 \times AMW \quad n = 48, R^2_{adj} = 0.95, RMSE = 0.85, F = 166.24, p < 0.001 \tag{15}$$

For PVA nanoplastics,

$$C_m = 8.694 - 0.018 \times AATS5m + 0.739 \times MATS8c + 0.682 \times GATS8c - 10.196 \times VE1\_DzZ + 1.676 \times AMW \quad n = 48, R^2_{adj} = 0.84, RMSE = 2.44, F = 48.61, p < 0.001 \tag{16}$$

For PMMA nanoplastics,

$$C_m = 4.009 - 0.01 \times AATS5m + 0.174 \times MATS8c + 0.242 \times GATS8c - 2.58 \times VE1\_DzZ + 0.825 \times AMW \quad n = 48, R^2_{adj} = 0.95, RMSE = 0.62, F = 165.85, p < 0.001 \tag{17}$$

All 12 QSAR models achieved R^2_{adj} values exceeding the criterion of $R^2 > 0.60$, indicating their strong goodness-of-fit. Moreover, no serious multi-collinearity was observed among the five variables, as their *VIF* values were below than 10 (Table S4). As shown in Figure 5, the predicted C_m values by the established QSAR models agree well with those obtained from GCMC simulations, further validating the performance of the models.

As illustrated in Figures S2 and S3, we also evaluated the ADs of the 12 QSAR models for C_m values using the δ^* and h values. Note that for these 12 QSAR models, the h values for fluorene (0.78) and heptabromodiphenyl ether (0.91) exceeded the warning leverage value h^* (0.375), while their δ^* values remained within the range of −3~3. These results indicate that the structures of fluorene and heptabromodiphenyl ether are significantly different from other compounds in the dataset, and they are influential on the QSAR models.

https://doi.org/10.3390/nano16030178

Additionally, some outliers were identified based on $|\delta^*|$ values (larger than 3): for the QSAR models of C_m values on PA, PC, PP, POM, PVA, and PMMA nanoplastics, formaldehyde is an outlier. For the QSAR model on PE, n-propylbenzene is an outlier. Likewise, 4-fluorophenol is an outlier for the QSAR model on PS nanoplastics.

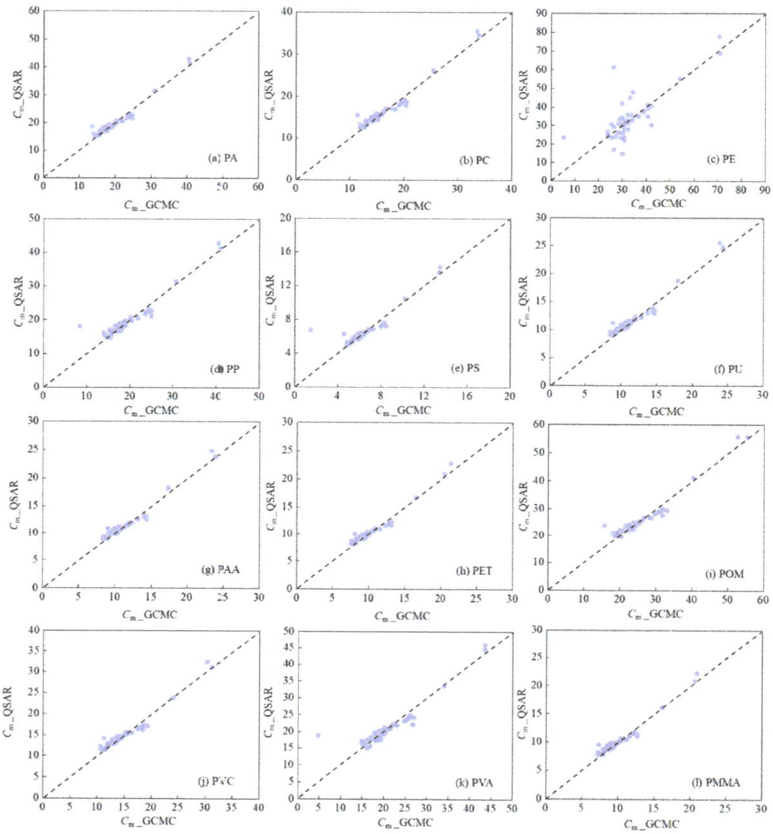

Figure 5. Predicted C_m values on (**a**) PA, (**b**) PC, (**c**) PE, (**d**) PP, (**e**) PS, (**f**) PU, (**g**) PAA, (**h**) PET, (**i**) POM, (**j**) PVC, (**k**) PVA and (**l**) PMMA with QSAR models (C_m_QSAR) versus those from GCMC simulations (C_m_GCMC).

In general, the AD of a prediction model depends on the dataset used. Since all 12 QSAR models were developed with the same dataset consisting of 48 different organic compounds, they share the same ADs covering a wide range of compounds with diverse functional groups, including -OH, $-NH_2$, $-NO_2$, $-CH_3$, -CN, -CHO, $-CH_2OH$, $-CH_2CH_3$, $-C(O)CH_3$, $-CH_2CH_2OH$, $-OC(O)CH_3$, $-C(O)OCH_3$, $-CH_2CH_2CH_3$, $-C(O)OCH_2CH_3$, -F, -Br, $OP-(OR)_3$, and $-C_6H_5$.

3.3.2. A Multi-Dimensional Prediction Model

Both the QSAR models for log$K$ values and those for C_m values developed above are specific to individual nanoplastics, as each model solely employs molecular structural descriptors of organic compounds as input features. These descriptors cannot capture the structural characteristics of the nanoplastics themselves, and thus are limited to single adsorbents. To overcome this limitation, we incorporated structural descriptors of the

 https://doi.org/10.3390/nano16030178

nanoplastics into the QSAR models, aiming at developing a multi-dimensional prediction model that estimate the adsorption of various organic compounds on a diverse range of nanoplastics.

Based on the 12 sorbent-specific QSAR models developed for predicting C_m values, we derived a simplified model to unify their predictive capabilities into a single framework:

$$C_m = c + x_1 \times AATS5m + x_2 \times MATS8c + x_3 \times GATS8c + x_4 \times VE1\_DzZ + x_5 \times AMW \quad (18)$$

where *AATS5m*, *MATS8c*, *GATS8c*, *VE1_DzZ*, and *AMW* are descriptors for the organic compounds. Recognizing that the adsorption is influenced not only by the structures of the organic compounds but also by the characteristics of nanoplastics, we made an attempt to incorporate some descriptors reflecting the structural features of nanoplastics. This approach allows us to predict the constant c and other coefficients (x_1, x_2, x_3, x_4, and x_5) in Equation (18), resulting in a prediction model with variables combining the structural information from both organic compounds and nanoplastics. The values for c, x_1, x_2, x_3, x_4, and x_5 can be estimated with the following equations.

$$c = 11.361 - 5.839 \times GATS3v, r = 0.889 \quad (19)$$

$$x_1 = -0.019 + 0.008 \times GATS3v, r = 0.787 \quad (20)$$

$$x_2 = 71.408 - 92.984 \times SIC2, r = 0.755 \quad (21)$$

$$x_3 = -24.086 + 31.432 \times SIC2, r = 0.749 \quad (22)$$

$$x_4 = -36.628 + 46.548 \times BIC3, r = 0.798 \quad (23)$$

$$x_5 = 2.305 - 1.242 \times GATS3v, r = 0.866 \quad (24)$$

Accordingly, the optimal multi-dimensional prediction model is shown below.

$$\begin{aligned} C_m = {} & (11.361 - 5.839 \times GATS3v) + (-0.019 + 0.008 \times GATS3v) \times AATS5m \\ & + (71.408 - 92.984 \times SIC2) \times MATS8c + (-24.086 + 31.432 \times SIC2) \times GATS8c + \\ & (-36.628 + 46.548 \times BIC3) \times VE1\_DzZ + (2.305 - 1.242 \times GATS3v) \times AMW \end{aligned} \quad (25)$$

Note that the values of x_2 and x_3 for PE differ significantly from those for other nanoplastics (Figure S4), making them influential in the univariate predictive model (Equations (21) and (22)). This observation highlights the need for future investigation into adsorption on a broader range of nanoplastics to expand the dataset. By incorporating more diverse nanoplastics and utilizing additional descriptors to better characterize their structural features, the robustness of the multi-dimensional prediction model can be enhanced, and its applicability domains (ADs) can be further expanded.

3.4. Adsorption Mechanisms

As presented in Equations (3)–(16) and Table S5, the descriptors utilized in the developed QSAR models capture various structural characteristics. To compare the contributions of these descriptors, we used radar compass plots (Figure 6) with standardized coefficients. Besides, the details for the standardized coefficients, t, p values, and *VIF* values of these descriptors, are given in Table S4. Note that the descriptors used for predicting the log$K$

https://doi.org/10.3390/nano16030178

values are different from those for predicting the C_m values (Table S4), suggesting that the factors influencing log*K* are different from those influencing C_m. This distinction offers us valuable insights into the underlying adsorption mechanisms driving adsorption on nanoplastics.

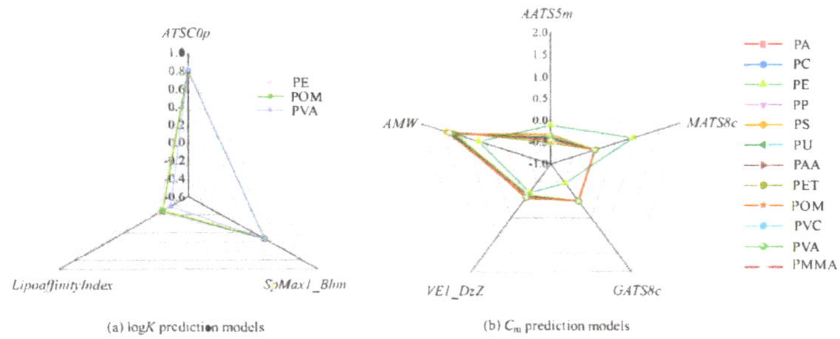

Figure 6. Radar compass plot with the standardized coefficients of (**a**) three descriptors in the log*K* prediction models and (**b**) five descriptors in the C_m prediction models.

3.4.1. QSAR Models for log*K* Values

As shown in Figure 6a, the standardized coefficients of three descriptors, *ATSC0p*, *SpMax1_Bhm*, and *LipoaffinityIndex*, in the QSAR models (Equations (3)–(5)) for predicting the log*K* values on PE, POM, and PVA nanoplastics are significantly different from each other. This variation implies that these descriptors exert distinct influences on adsorption, reflecting the unique interactions between the nanoplastics and the organic pollutants.

For the adsorption on PE, POM, and PVA nanoplastics, the standardized coefficients of *ATSC0p* are significantly higher than those of *SpMax1_Bhm* and *LipoaffinityIndex* [Figure 6a]. It indicates that *ATSC0p* is the most influential predictive variable in predicting the log*K* values for organic compounds on PE, POM and PVA nanoplastics. The positive standardized coefficients of *ATSC0p* suggest that an increase in this descriptor for an organic compound can lead to an increase in the log*K* values.

ATSC0p [63] represents the centered Broto–Moreau autocorrelation-lag0/weighted by polarizabilities. It means that the polarizabilities of organic compounds can influence their adsorption on nanoplastics. Note that compounds with higher polarizability tend to have larger *ATSC0p* values. For example, benzene has a polarizability of 10.32, lower than that of cyclohexane (11.0) [64], and a correspondingly smaller *ATSC0p* value (3.02 for benzene vs. 4.01 for cyclohexane). Accordingly, the log*K* values for benzene on PE (5.74) and POM (6.13) are lower than those for cyclohexane (6.94 on PE and 8.62 on POM). However, for the adsorption on PVA, the log*K* value for benzene (5.76) is slightly higher than that for cyclohexane (5.13). This trend suggests that compounds with higher polarizability are more prone to be adsorbed by PE and POM nanoplastics, which may be attributed to stronger induced and dispersion forces between compounds with higher polarizability and these nanoplastics, leading to higher log*K* values on the nanoplastics.

Additionally, the standardized coefficients of *SpMax1_Bhm* in the QSAR models for predicting the log*K* values on PE, POM, and PVA nanoplastics are all positive, while those of *LipoaffinityIndex* are negative. Thus, the log*K* values can increase with increasing *SpMax1_Bhm* values for the organic compounds, while a decrease in *LipoaffinityIndex* also leads to higher log*K* values.

SpMax1_Bhm [63] represents the largest absolute eigenvalue of the Burden modified matrix-n1 weighted by relative mass. In general, molecules with larger mass tend to have higher molar volumes, and consequently more contact with nanoplastics, thereby having stronger van der Waals interactions. In contrast, the descriptor *LipoaffinityIndex* [65], which

 https://doi.org/10.3390/nano16030178

measures the lipoaffinity of an organic compound, is negatively correlated with log*K* values. Compounds having functional groups such as -OH, -NO_2, and -NH_2 typically have lower *LipoaffinityIndex* values, making them more readily adsorbed by nanoplastics. This may be because these functional groups can enhance dipole-induced dipole interactions between the organic pollutants and nanoplastics, thereby increasing the log*K* values.

3.4.2. Sorbent-Specific QSAR Models for C_m Values

As shown in Equations (6)–(17), five descriptors, namely, *AATS5m*, *MATS8c*, *GATS8c*, *VE1_DzZ*, and *AMW*, are used in combination to predict the C_m values on 12 nanoplastics (PA, PC, PE, PP, PS, PU, PAA, PET, POM, PVC, PVA, and PMMA). Figure 6b illustrates that the standardized coefficients for these five descriptors vary in magnitude, highlighting their different contributions to predicting the C_m values.

Among these five descriptors, *AMW* exhibits the highest standardized coefficients in the QSAR models for predicting C_m values on all nanoplastics except PE, indicating that *AMW* is the most influential parameter in these predictions. *AMW* [66] denotes the average molecular weight (molecular weight divided by the total number of atoms). The positive standardized coefficients for *AMW* suggest that higher C_m values are associated with increased *AMW* values of organic compounds. In contrast, the standardized coefficients for *AATS5m* are negative, which implies that the C_m values increase as *AATS5m* decreases. *AATS5m* [67] represents average Broto–Moreau autocorrelation-lag5 weighted by mass, which is also related to the molecular weight. These findings highlight the critical role of molecular weight, as characterized by *AMW* and *AATS5m*, in predicting the C_m values.

Additionally, the standardized coefficients for *MATS8c* and *GATS8c* are positive in the QSAR models for predicting C_m values on all nanoplastics except PE, which indicates that the C_m values increase with higher *MATS8c* and *GATS8c* values. *MATS8c* [68] is Moran autocorrelation-lag8 weighted by charges, while *GATS8c* [69] represents Geary autocorrelation-lag8 weighted by charges. These two descriptors capture charge-related interactions well, suggesting that compounds with higher charges (reflected by higher *MATS8c* and *GATS8c* values) exhibit stronger electrostatic interactions with nanoplastics, thus leading to higher C_m values.

Lastly, the descriptor *VE1_DzZ* [63], which represents the coefficient sum of the last eigenvector from Barysz matrix weighted by atomic number, has negative standardized coefficients, which indicates that *VE1_DzZ* contributes negatively to the C_m values.

Notably, on PE nanoplastics, the standardized coefficient of *GATS8c* for predicting the C_m values is negative, in contrast to its positive correlation on the other 11 nanoplastics. The negative coefficient indicates that, for PE, *GATS8c* is negatively correlated with C_m values. For PE nanoplastics, *MATS8c* has the highest standardized coefficient, indicating that it is the most influential parameter among these five descriptors in predicting C_m values. Similar to their role in the other 11 nanoplastics, *GATS8c* and *MATS8c* combinedly reflect charge-related interactions, which contributes to the prediction of C_m values on PE nanoplastics.

3.4.3. Multi-Dimensional QSAR Models for C_m Values

According to the multi-dimensional prediction model (Equation (25)), three additional descriptors, namely, *GATS3v*, *SIC2*, and *BIC3*, characterizing the structural properties of the nanoplastics, also influence C_m values.

GATS3v [70] represents Geary autocorrelation-lag3 weighted by van der Waals volumes, highlighting the significant role of van der Waals interactions in the adsorption of organic compounds on nanoplastics. It is consistent with the adsorption mechanisms revealed by the QSAR models for predicting log*K* values.

https://doi.org/10.3390/nano16030178

In addition, the descriptors *SIC2* and *BIC3* are also related to the prediction of C_m values. *SIC2* [63] denotes structural information content index (neighborhood symmetry of 2-order), while *BIC3* [71] is bond information content index (neighborhood symmetry of 3-order).

3.4.4. Effects of Functional Groups on Adsorption Energies

To explore the effects of functional groups on adsorption, we compared the adsorption energies of compounds with different functional groups on PE, POM, and PVA nanoplastics (Table S6). Note that the adsorption energies for benzene on PE (−7.36 kcal/mol) and POM (−8.54 kcal/mol) are more negative than those for cyclohexane on PE (−6.37 kcal/mol) and POM (−8.18 kcal/mol). This difference may arise from the π electrons in benzene, which enhance the electrostatic interactions with PE and POM nanoplastics. In contrast, the adsorption energy for benzene on PVA (−7.44 kcal/mol) is higher than that for cyclohexane (−8.52 kcal/mol), likely due to repulsive interactions between the π electrons of benzene and the electrons of -OH group of PVA, which reduce the adsorption.

For benzene derivatives with mono-substituent, the interactions with PE/POM/PVA nanoplastics are generally stronger than those with benzene, irrespective of the substituent's electrophilic properties. For example, benzyl alcohol, with a substituent ($-CH_2OH$) that has a Hammett sigma meta parameter of zero ($\sigma_m = 0$), exhibits stronger adsorption energies on PE (−9.79 kcal/mol), POM (−14.81 kcal/mol), and PVA (−16.11 kcal/mol) compared to benzene on these nanoplastics (PE, −7.36 kcal/mol; POM, −8.54 kcal/mol; PVA, −7.44 kcal/mol). The enhanced adsorption can be ascribed to the dispersion interactions between the substituents and the nanoplastics, which promote stronger binding of benzene derivatives.

Notably, nitrobenzene exhibits slightly weaker adsorption energy on POM compared to benzene, which could be due to electrostatic repulsion between the electron-withdrawing $-NO_2$ group on nitrobenzene and the electron-rich oxygen atoms in POM, weakening the adsorption.

4. Conclusions

Based on GCMC simulations, PE, POM, and PVA nanoplastics exhibit the highest adsorption capacities among the twelve nanoplastics investigated for the 39 representative aliphatic and aromatic compounds and the nine emerging pollutants (brominated flame retardants and phosphorus flame retardants). The QSAR analysis further reveals that *ATSC0p*, a descriptor associated with molecular polarizability, is the most influential factor governing log*K* values on PE, POM, and PVA, underscoring the key role of dispersion-related effects in pollutant uptake. For adsorption capacities (C_m), a set of QSAR models was established for all twelve nanoplastics, and an integrated multi-dimensional prediction model was further developed to enable rapid, one-step screening of adsorption capacities for diverse pollutants across multiple nanoplastic types.

Mechanistically, all these developed models consistently indicate that van der Waals interactions, supplemented by electrostatic interactions for more polar compounds, dominate adsorption of organic compounds on nanoplastics under gas-phase conditions relevant to atmospheric environments. Collectively, the proposed GCMC-DFT-QSAR framework and the multi-dimensional predictive model substantially improve the efficiency of generating adsorption data for pollutant–nanoplastic pairs, providing a practical basis for screening environmental behavior and supporting ecological risk assessment of organic pollutants associated with atmospheric nanoplastics.

Supplementary Materials: The following supporting information can be downloaded at: https://www.mdpi.com/article/10.3390/nano16030178/s1, Table S1: Organic compounds and estimated

https://doi.org/10.3390/nano16030178

adsorption capacity (C_m, g/g) from GCMC simulations and logarithm values for adsorption equilibrium coefficient (log*K*) from DFT computations under atmospheric conditions; Table S2: The atmospheric log*K* values for five organic compounds on PE NPs from DFT computations and those from MD simulations; Table S3: Adsorption equilibrium configuration for the 48 organic compounds towards PE, POM and PVA nanoplastics; Table S4: Standardized coefficients, *t*, *p* values and variable inflation factor (*VIF*) for the predictive variables; Table S5: Definitions for the descriptors utilized in the developed models; Table S6: Hammett parameters of substituents for 10 different organic compounds and their adsorption energies (E_{ad}) on PE, POM and PVA nanoplastics; Figure S1: Application domain characterized by Williams plots with standardized residuals (δ^*) and leverage values (*h*) for QSAR models of log*K* values on (a) PE, (b) POM and (c) PVA nanoplastics (h^*: the warning leverage value); Figure S2: Application domain characterized by Williams plots with standardized residuals (δ^*) and leverage values (*h*) for QSAR models of C_m values on (a) PA, (b) PC, (c) PE, (d) PP, (e) PS and (f) PU nanoplastics (h^*: the warning leverage value); Figure S3: Application domain characterized by Williams plots with standardized residuals (δ^*) and leverage values (*h*) for QSAR models of C_m values on (a) PAA, (b) PET, (c) POM, (d) PVC, (e) PVA and (f) PMMA nanoplastics (h^*: the warning leverage value); Figure S4: Prediction for c, x_1, x_2, x_3, x_4 and x_5 based on descriptors characterizing nanoplastics.

Author Contributions: The manuscript was written through the contributions of all authors. Conceptualization, Y.W., H.Y. and X.T.; methodology, Y.W., C.L. and Z.C.; software, Y.W., C.L. and Z.C.; validation, Y.W. and P.Z.; formal analysis, Y.W., C.L. and Z.C.; investigation, Y.W.; data curation, Y.W., H.Y., X.T. and P.Z.; writing—original draft preparation, Y.W.; writing—review and editing, Y.W., H.Y., C.L., X.T., P.Z. and Z.C.; visualization, Y.W. and P.Z.; supervision, Y.W.; funding acquisition, Y.W., X.T. and H.Y. All authors have read and agreed to the published version of the manuscript.

Funding: This research was funded by the National Natural Science Foundation of China, grant number 22206095; the Fundamental Research Funds for the Central Universities, grant number FRF-TP-22-087A1 and FRF-BD-25-003; and the National Key R&D Program of China, grant number 2023YFB3810800.

Data Availability Statement: The data supporting this article have been included in the Supplementary Information.

Conflicts of Interest: The authors declare no competing financial interest.

References

1. Geyer, R.; Jambeck, J.R.; Law, K.L. Production, use, and fate of all plastics ever made. *Sci. Adv.* **2017**, *3*, e1700782. [CrossRef] [PubMed]
2. Annual Production of Plastics Worldwide from 1950 to 2022. Available online: https://www.statista.com/statistics/282732/global-production-of-plastics-since-1950/ (accessed on 15 December 2025).
3. Stegmann, P.; Daioglou, V.; Londo, M.; van Vuuren, D.P.; Junginger, M. Plastic futures and their CO_2 emissions. *Nature* **2022**, *612*, 272–276. [CrossRef]
4. Alimi, O.S.; Budarz, J.F.; Hernandez, L.M.; Tufenkji, N. Microplastics and nanoplastics in aquatic environments: Aggregation, deposition, and enhanced contaminant transport. *Environ. Sci. Technol.* **2018**, *52*, 1704–1724. [CrossRef]
5. Botterell, Z.L.R.; Beaumont, N.; Dorrington, T.; Steinke, M.; Thompson, R.C.; Lindeque, P.K. Bioavailability and effects of microplastics on marine zooplankton: A review. *Environ. Pollut.* **2019**, *245*, 98–110. [CrossRef]
6. Macleod, M.; Arp, H.P.H.; Tekman, M.B.; Jahnke, A. The global threat from plastic pollution. *Science* **2021**, *373*, 61–65. [CrossRef]
7. Thompson, R.C.; Olsen, Y.; Mitchell, R.P.; Davis, A.; Rowland, S.J.; John, A.W.G.; McGonigle, D.; Russell, A.E. Lost at sea: Where is all the plastic? *Science* **2004**, *304*, 838. [CrossRef]
8. Hochella, M.F.; Mogk, D.W.; Ranville, J.; Allen, I.C.; Luther, G.W.; Marr, L.C.; McGrail, B.P.; Murayama, M.; Qafoku, N.P.; Rosso, K.M.; et al. Natural, incidental, and engineered nanomaterials and their impacts on the Earth system. *Science* **2019**, *363*, 1414. [CrossRef]
9. Rainieri, S.; Conlledo, N.; Larsen, B.K.; Granby, K.; Barranco, A. Combined effects of microplastics and chemical contaminants on the organ toxicity of zebrafish (*Danio rerio*). *Environ. Res.* **2018**, *162*, 135–143. [CrossRef]
10. Xiong, Y.C.; Zhao, J.H.; Li, L.Q.; Wang, Y.Y.; Dai, X.H.; Yu, F.; Ma, J. Interfacial interaction between micro/nanoplastics and typical PPCPs and nanoplastics removal via electrosorption from an aqueous solution. *Water Res.* **2020**, *184*, 116100. [CrossRef] [PubMed]

https://doi.org/10.3390/nano16030178

11. Yang, W.F.; Gao, X.X.; Wu, Y.X.; Wan, L.; Tan, L.C.; Yuan, S.M.; Ding, H.J.; Zhang, W.H. The combined toxicity influence of microplastics and nonylphenol on microalgae *Chlorella pyrenoidosa*. *Ecotoxicol. Environ. Saf.* **2020**, *195*, 110484. [CrossRef] [PubMed]
12. Wu, C.X.; Zhang, K.; Huang, X.L.; Liu, J.T. Sorption of pharmaceuticals and personal care products to polyethylene debris. *Environ. Sci. Pollut. Res.* **2016**, *23*, 8819–8826. [CrossRef]
13. Velzeboer, I.; Kwadijk, C.; Koelmans, A.A. Strong sorption of PCBs to nanoplastics, microplastics, carbon nanotubes, and fullerenes. *Environ. Sci. Technol.* **2014**, *48*, 4869–4876. [CrossRef]
14. Bhagwat, G.; Tran, T.; Lamb, D.; Senathirajah, K.; Grainge, I.; O'Connor, W.; Juhasz, A.; Palanisami, T. Biofilms enhance the adsorption of toxic contaminants on plastic microfibers under environmentally relevant conditions. *Environ. Sci. Technol.* **2021**, *55*, 8877–8887. [CrossRef] [PubMed]
15. Ding, L.; Mao, R.F.; Ma, S.R.; Guo, X.T.; Zhu, L.Y. High temperature depended on the ageing mechanism of microplastics under different environmental conditions and its effect on the distribution of organic pollutants. *Water Res.* **2020**, *174*, 115634. [CrossRef]
16. Dong, X.F.; Zheng, M.G.; Qu, L.Y.; Shi, L.; Wang, L.; Zhang, Y.; Liu, X.W.; Qiu, Y.; Zhu, H.H. Sorption of tonalide, musk xylene, galaxolide, and musk ketone by microplastics of polyethylene and polyvinyl chloride. *Mar. Pollut. Bull.* **2019**, *144*, 129–133. [CrossRef] [PubMed]
17. Fu, L.N.; Li, J.; Wang, G.Y.; Luan, Y.N.; Dai, W. Adsorption behavior of organic pollutants on microplastics. *Ecotoxicol. Environ. Saf.* **2021**, *217*, 112207. [CrossRef] [PubMed]
18. Liu, X.M.; Shi, H.H.; Xie, B.; Dionysiou, D.D.; Zhao, Y.P. Microplastics as both a sink and a source of bisphenol A in the marine environment. *Environ. Sci. Technol.* **2019**, *53*, 10188–10196. [CrossRef]
19. Liu, G.; Zhu, Z.; Yang, Y.; Sun, Y.; Yu, F.; Ma, J. Sorption behavior and mechanism of hydrophilic organic chemicals to virgin and aged microplastics in freshwater and seawater. *Environ. Pollut.* **2019**, *246*, 26–33. [CrossRef]
20. Liu, P.; Lu, K.; Li, J.; Wu, X.; Qian, L.; Wang, M.; Gao, S. Effect of aging on adsorption behavior of polystyrene microplastics for pharmaceuticals: Adsorption mechanism and role of aging intermediates. *J. Hazard. Mater.* **2020**, *384*, 121193. [CrossRef]
21. Wang, F.; Shih, K.M.; Li, X.Y. The partition behavior of perfluorooctanesulfonate (PFOS) and perfluorooctanesulfonamide (FOSA) on microplastics. *Chemosphere* **2015**, *119*, 841–847. [CrossRef]
22. Schäfer, A.I.; Nghiem, L.D.; Waite, T.D. Removal of the natural hormone estrone from aqueous solutions using nanofiltration and reverse osmosis. *Environ. Sci. Technol.* **2002**, *37*, 182–188. [CrossRef]
23. Liu, X.; Xu, J.; Zhao, Y.; Shi, H.; Huang, C.H. Hydrophobic sorption behaviors of 17β-Estradiol on environmental microplastics. *Chemosphere* **2019**, *226*, 726–735. [CrossRef]
24. Louie, S.G.; Chan, Y.H.; da Jornada, F.H.; Li, Z.L.; Qiu, D.Y. Discovering and understanding materials through computation. *Nat. Mater.* **2021**, *20*, 728–735. [CrossRef] [PubMed]
25. Mo, Q.M.; Yang, X.J.; Wang, J.J.; Xu, H.J.; Li, W.Y.; Fan, Q.; Gao, S.; Yang, W.Y.; Gao, C.Z.; Liao, D.H.; et al. Adsorption mechanism of two pesticides on polyethylene and polypropylene microplastics: DFT calculations and particle size effects. *Environ. Pollut.* **2021**, *291*, 118120. [CrossRef] [PubMed]
26. Enyoh, C.E.; Wang, Q.; Wang, W.; Chowdhury, T.; Rabin, M.H.; Islam, R.; Yue, G.; Lin, Y.; Xiao, K. Sorption of per- and polyfluoroalkyl substances (PFAS) using polyethylene (PE) microplastics as adsorbent: Grand canonical Monte Carlo and molecular dynamics (GCMC-MD) studies. *Int. J. Environ. Anal. Chem.* **2022**, *104*, 2719–2735. [CrossRef]
27. Joliat, J.; Picaud, S.; Jedlovszky, P. Competitive adsorption of trace gases on ice at tropospheric temperatures: A grand canonical Monte Carlo simulation study. *J. Phys. Chem. A* **2023**, *127*, 10223–10232. [CrossRef]
28. Sun, Y.X.; Ma, Y.J.; Yu, B.; Zhang, W.; Zhang, L.P.; Chen, P.; Xu, L. Perspectives on molecular simulation of CO_2/CH_4 competitive adsorption in a shale matrix: A review. *Energy Fuels* **2024**, *38*, 15935–15971. [CrossRef]
29. Cai, S.Y.; Yu, L.; Huo, E.G.; Ren, Y.X.; Liu, X.D.; Chen, Y.P. Adsorption and diffusion properties of functionalized MOFs for CO_2 capture: A combination of molecular dynamics simulation and density functional theory calculation. *Langmuir* **2024**, *40*, 6869–6877. [CrossRef]
30. Xu, J.H.; Carney, T.E.; Zhou, R.Y.; Shepard, C.; Kanai, Y. Real-time time-dependent density functional theory for simulating nonequilibrium electron dynamics. *J. Am. Chem. Soc.* **2024**, *146*, 5011–5029. [CrossRef]
31. Li, M.; Yu, H.Y.; Wang, Y.F.; Li, J.G.; Ma, G.C.; Wei, X.X. QSPR models for predicting the adsorption capacity for microplastics of polyethylene, polypropylene and polystyrene. *Sci. Rep.* **2020**, *10*, 14597. [CrossRef] [PubMed]
32. Gui, B.X.; Xu, X.T.; Zhang, S.N.; Wang, Y.; Li, C.; Zhang, D.M.; Su, L.M.; Zhao, Y.H. Prediction of organic compounds adsorbed by polyethylene and chlorinated polyethylene microplastics in freshwater using QSAR. *Environ. Res.* **2021**, *197*, 111001. [CrossRef]
33. Xu, J.P.; Wang, L.; Sun, H.W. Adsorption of neutral organic compounds on polar and nonpolar microplastics: Prediction and insight into mechanisms based on pp-LFERs. *J. Hazard. Mater.* **2021**, *408*, 124857. [CrossRef]
34. Yao, J.J.; Wen, J.Y.; Li, H.P.; Yang, Y. Surface functional groups determine adsorption of pharmaceuticals and personal care products on polypropylene microplastics. *J. Hazard. Mater.* **2022**, *423*, 127131. [CrossRef]

35. Qiu, Y.; Li, Z.J.; Zhang, T.; Zhang, P. Predicting aqueous sorption of organic pollutants on microplastics with machine learning. *Water Res.* **2023**, *244*, 120503. [CrossRef]
36. Akkermans, R.L.; Spenley, N.A.; Robertson, S.H. Monte Carlo methods in materials studio. *Mol. Simul.* **2013**, *39*, 1153–1164. [CrossRef]
37. Nosé, S. A molecular dynamics method for simulations in the canonical ensemble. *Mol. Phys.* **1984**, *52*, 255–268. [CrossRef]
38. Sun, H.; Jin, Z.; Yang, C.W.; Akkermans, R.L.C.; Robertson, S.H.; Spenley, N.A.; Miller, S.; Todd, S.M. COMPASS II: Extended coverage for polymer and drug-like molecule databases. *J. Mol. Model.* **2016**, *22*, 47. [CrossRef]
39. Sun, H. COMPASS: An ab initio forcefield optimized for condensed-phase applications—Overview with details on alkane and benzene compounds. *J. Phys. Chem. B* **1998**, *102*, 7338–7364. [CrossRef]
40. Sun, H.; Ren, P.; Fried, J.R. The COMPASS forcefield: Parameterization and validation for polyphosphazenes. *Comput. Theor. Polym. Sci.* **1998**, *8*, 229–246. [CrossRef]
41. Rigby, D.; Sun, H.; Eichinger, B.E. Computer simulations of poly(ethylene oxides): Forcefield, PVT diagram and cyclization behavior. *Polym. Int.* **1997**, *44*, 311–330. [CrossRef]
42. Wells, B.A.; Chaffee, A.L. Ewald summation for molecular simulations. *J. Chem. Theory Comput.* **2015**, *11*, 3684–3695. [CrossRef]
43. Brooks, C.L., III; Montgomery Pettitt, B.; Karplus, M. Structural and energetic effects of truncating long range interactions in ionic and polar fluids. *J. Chem. Phys.* **1985**, *83*, 5897–5908. [CrossRef]
44. Ni, B.; Baumketner, A. Effect of atom- and group-based truncations on biomolecules simulated with reaction-field electrostatics. *J. Mol. Model.* **2011**, *17*, 2883–2893. [CrossRef] [PubMed]
45. Putri, A.D.; Murti, B.T.; Kanchi, S.; Sabela, M.I.; Bisetty, K.; Tiwari, A.; Inamuddin; Asiri, A.M. Computational studies on the molecular insights of aptamer induced poly(N-isopropylacrylamide)-graft-graphene oxide for on/off- switchable whole-cell cancer diagnostics. *Sci. Rep.* **2019**, *9*, 7873. [CrossRef] [PubMed]
46. Nyden, M.R.; Stoliarov, S.I.; Westmoreland, P.R.; Guo, Z.X.; Jee, C. Applications of reactive molecular dynamics to the study of the thermal decomposition of polymers and nanoscale structures. *Mat. Sci. Eng. A* **2004**, *365*, 114–121. [CrossRef]
47. Delley, B. An all-electron numerical method for solving the local density functional for polyatomic molecules. *J. Chem. Phys.* **1990**, *92*, 508–517. [CrossRef]
48. Delley, B. From molecules to solids with the $DMol^3$ approach. *J. Chem. Phys.* **2000**, *113*, 7756–7764. [CrossRef]
49. Perdew, J.P.; Burke, K.; Ernzerhof, M. Generalized gradient approximation made simple. *Phys. Rev. Lett.* **1996**, *77*, 3865–3868. [CrossRef]
50. Benedek, N.A.; Snook, I.K.; Latham, K.; Yarovsky, I. Application of numerical basis sets to hydrogen bonded systems: A density functional theory study. *J. Chem. Phys.* **2005**, *122*, 144102. [CrossRef] [PubMed]
51. Inada, Y.; Orita, H. Efficiency of numerical basis sets for predicting the binding energies of hydrogen bonded complexes: Evidence of small basis set superposition error compared to Gaussian basis sets. *J. Comput. Chem.* **2008**, *29*, 225–232. [CrossRef]
52. Liu, P.; Rodriguez, J.A. Catalysts for hydrogen evolution from the [NiFe] hydrogenase to the Ni_2P (001) surface: The importance of ensemble effect. *J. Am. Chem. Soc.* **2005**, *127*, 14871–14878. [CrossRef]
53. Grimme, S. Semiempirical GGA-type density functional constructed with a long-range dispersion correction. *J. Comput. Chem.* **2006**, *27*, 1787–1799. [CrossRef]
54. Zou, M.Y.; Zhang, J.D.; Chen, J.W.; Li, X.H. Simulating adsorption of organic pollutants on finite (8,0) single-walled carbon nanotubes in water. *Environ. Sci. Technol.* **2012**, *46*, 8887–8894. [CrossRef]
55. Hong, Q.Y.; Xu, H.M.; Ling, C.C.; Liu, Z.S.; Huang, W.J.; Yang, S.J.; Qu, Z.; Zhang, L.Z.; Yan, N.Q. Low-coordinated polysulfide species on metal sulfide enabled superior gaseous mercury removal from natural gas. *Environ. Sci. Technol.* **2025**, *59*, 4178–4187. [CrossRef]
56. Yap, C.W. PaDEL-Descriptor: An open source software to calculate molecular descriptors and fingerprints. *J. Comput. Chem.* **2011**, *32*, 1466–1474. [CrossRef]
57. Gramatica, P. Principles of QSAR models validation: Internal and external. *QSAR Comb. Sci.* **2007**, *26*, 694–701. [CrossRef]
58. Wang, Y.; Chen, J.W.; Wei, X.X.; Hernandez Maldonado, A.J.; Chen, Z.F. Unveiling adsorption mechanisms of organic pollutants onto carbon nanomaterials by density functional theory computations and linear free energy relationship modeling. *Environ. Sci. Technol.* **2017**, *51*, 11820–11828. [CrossRef] [PubMed]
59. Wang, Y.; Tang, W.H.; Peng, Y.; Chen, Z.F.; Chen, J.W.; Xiao, Z.J.; Zhao, X.G.; Qu, Y.K.; Li, J.H. Predicting the adsorption of organic pollutants on boron nitride nanosheets *via in silico* techniques: DFT computations and QSAR modeling. *Environ. Sci. Nano* **2021**, *8*, 795–805. [CrossRef]
60. Su, L.H.; Wang, Z.Y.; Xiao, Z.J.; Xia, D.M.; Wang, Y.; Chen, J.W. Rapidly predicting aqueous adsorption constants of organic pollutants onto polyethylene microplastics by combining molecular dynamics simulations and machine learning. *ACS EST Water* **2024**, *4*, 4184–4192. [CrossRef]
61. Golbraikh, A.; Shen, M.; Xiao, Z.Y.; Xiao, Y.D.; Lee, K.H.; Tropsha, A. Rational selection of training and test sets for the development of validated QSAR models. *J. Comput. Aided Mol. Des.* **2003**, *17*, 241–253. [CrossRef] [PubMed]

62. Robinson, C.; Schumacker, R.E. Interaction effects: Centering, variance inflation factor, and interpretation issues. *Mult. Linear Regres. Viewp.* **2009**, *35*, 6–11.
63. Todeschini, R.; Consonni, V. *Molecular Descriptors for Chemoinformatics*; Wiley VCH: Weinheim, Germany, 2009; pp. 27–37.
64. Lide, D.R. *CRC Handbook of Chemistry and Physics*; CRC Press, Inc.: Boca Raton, FL, USA, 2005.
65. Hall, L.H.; Kier, L.B. Electrotopological state indices for atom types: A novel combination of electronic, topological, and valence state information. *J. Chem. Inf. Comput. Sci.* **1995**, *35*, 1039–1045. [CrossRef]
66. Deboyace, K.; Bookwala, M.; Buckner, I.S.; Zhou, D.; Wildfong, P.L.D. Interpreting the physicochemical meaning of a molecular descriptor predictive of amorphous solid dispersion formation in polyvinylpyrrolidone vinyl acetate. *Mol. Pharmaceut.* **2022**, *19*, 303–317. [CrossRef]
67. Prabhakar, Y.S.; Rawal, R.K.; Gupta, M.K.; Solomon, V.R.; Katti, S.B. Topological descriptors in modeling the HIV inhibitory activity of 2-aryl-3-pyridylthiazolidin-4-ones. *Comb. Chem. High T. Scr.* **2005**, *8*, 431–437.
68. Toots, K.M.; Sild, S.; Leis, J.; Acree, W.E.; Maran, U., Jr. Machine learning quantitative structure–property relationships as a function of ionic liquid cations for the gas-ionic liquid partition coefficient of hydrocarbons. *Int. J. Mol. Sci.* **2022**, *23*, 7534. [CrossRef] [PubMed]
69. Akinola, L.K.; Uzairu, A.; Shallangwa, G.A.; Abechi, S.E. Development and validation of predictive quantitative structure–activity relationship models for estrogenic activities of hydroxylated polychlorinated biphenyls. *Environ. Toxicol. Chem.* **2023**, *42*, 823–834. [CrossRef]
70. Euldji, I.; Si-Moussa, C.; Hamadache, M.; Benkortbi, O. QSPR modelling of the solubility of drug and drug-like compounds in supercritical carbon dioxide. *Mol. Inform.* **2022**, *41*, 2200026. [CrossRef] [PubMed]
71. Rezić, T.; Vrsalović Presečki, A.; Kurtanjek, Ž. New approach to the evaluation of lignocellulose derived by-products impact on lytic-polysaccharide monooxygenase activity using a molecular descriptor structural causality model. *Bioresour. Technol.* **2021**, *342*, 125990. [CrossRef] [PubMed]

https://doi.org/10.3390/nano16030178

nanomaterials

MDPI

Article

Interaction of Oxygen Molecules with Fe Atom-Doped γ-Graphyne Surfaces: First-Principles Calculations

Bin Zhao *, Jiayi Yin, Zhuoting Xiong, Wentao Yang, Peng Guo, Meng Li, Haoxian Zeng and Jianjun Wang *

Zhengzhou Key Laboratory of Low-Dimensional Quantum Materials and Devices, College of Physics and Optoelectronic Engineering, Zhongyuan University of Technology, Zhengzhou 450007, China
* Correspondence: zhaobin@whu.edu.cn (B.Z.); jjwang@zut.edu.cn (J.W.)

Abstract: The activation and dissociation of O_2 molecules play a key role in the oxidation of toxic gas molecules and the oxygen reduction reaction (ORR) in hydrogen–oxygen fuel cells. The interactions between O_2 molecules and the surfaces of Fe-doped γ-graphyne were systematically explored, mainly adopting the combined method of the density functional theory with dispersion correction (DFT-D3) and the climbing image nudged elastic band (CI-NEB) method. The order of the formation energy values of these defective systems is $E_f(Fe_{C2}) < E_f(Fe_{C1}) < E_f(Fe_{D1}) < E_f(V_{C1}) < E_f(V_{D1}) < E_f(V_{C2}) < E_f(Fe_{D2}) < E_f(V_{D2})$, which indicates that the process of Fe dopant atoms substituting single-carbon atoms/double-carbon atoms is relatively easier than the formation of vacancy-like defects. The results of ab initio molecular dynamics (AIMD) simulations confirm that the doped systems can maintain structural stability at room temperature conditions. Fe-doped atoms transfer a certain amount of electrons to the adsorbed O_2 molecules, thereby causing an increase in the O-O bond length of the adsorbed O_2 molecules. The electrons obtained by the anti-bonding $2\pi^*$ orbitals of the adsorbed O_2 molecules are mainly derived from the 3d orbitals of Fe atoms. There is a competitive relationship between the substrate's carbon atoms and the adsorbed O_2 molecules for the charges transferred from Fe atoms. In the C1 and C2 systems, O_2 molecules have a greater advantage in electron accepting ability compared to the substrate's carbon atoms. The elongation of O-O bonds and the amount of charge transfer exhibit a positive relationship. More electrons are transferred from Fe-3d orbitals to adsorbed O_2 molecules, occupying the $2\pi^*$ orbitals of adsorbed O_2 molecules, further elongating the O-O chemical bond until it breaks. The dissociation process of adsorbed O_2 molecules on the surfaces of GY-Fe systems (C2 and D2 sites) involves very low energy barriers (0.016 eV for C2 and 0.12 eV for D2). Thus, our studies may provide useful insights for designing catalyst materials for oxidation reactions and the oxygen reduction reaction.

Keywords: γ-graphyne; doped; oxygen activation; dissociation behavior; first-principles calculation

1. Introduction

Due to the many excellent properties of graphene materials, more and more researchers are paying attention to new carbon allotropes with a graphene-like structure. As a new two-dimensional material, γ-graphyne is identified as a direct bandgap semiconductor, and it is recognized to have the highest stability among the graphynes family [1–3]. In contrast to graphene, graphyne contains two types of hybridized carbon atoms, namely sp^- and sp^{2-} hybridized ones, which leads to the unique charge distribution on the surfaces [2–6].

Therefore, surface electronic state modulation can be achieved by doping different types of atoms, thus improving the surface chemical activity and endowing the material surface with peculiar physical and chemical properties [7–10]. The achievement of producing large-area, high-quality γ-graphyne has been realized recently through mechanochemistry, which has effectively promoted research on the properties and applications of graphyne [11,12]. Some studies indicate that graphynes have promising applications in hydrogen storage, gas sensors and catalysts [13–23]. Similarly to layered graphene, γ-graphyne is considered a potential catalyst for oxidation reactions and the oxygen reduction reaction (ORR). Catalytic activity, meanwhile, is dependent on the adsorption and dissociation of oxygen molecules, which means activated oxygen is vital to these reactions [24,25].

However, the large-area basal plane of γ-graphyne exhibits chemical inertness [26], and free O_2 molecules do not induce a sensitive response on its (001) surface. This necessitates the search for an effective strategy to activate the γ-graphyne basal plane for catalytic reactions. Theoretical and experimental studies have shown that iron/ruthenium–nitrogen co-doped carbon materials (Fe/Ru-N-C) are considered to be the most active non-precious metal ORR catalysts, with the FeN_4/RuN_4 coordination structure being the main catalytic active center [27–29]. Some theoretical research results have found that doped graphynes with different configurations have different catalytic performances for the oxygen reduction reaction (ORR). For example, α-graphyne with single B/N doping has low catalytic activity, but singly doped γ-graphyne exhibits certain catalytic activity for ORR [30]. Recently, Cui et al. reported that the anchored transition metal single atoms in nitrogen doped γ-graphyne can achieve efficient CO electrochemical reduction reaction (COER) [31]. Zheng et al. found that the single-atom Ti-doped graphdiyne is beneficial for the dissociation reaction of H_2O [5]. Our recent work also found that transition metal single-atom-doped γ-graphyne can achieve efficient H_2O dissociation [26]. Therefore, considering the easy doping characteristics of the graphyne surface, atomic doping can be used to modify the surface of graphyne, achieving a significant improvement in its catalytic performance.

In this work, we mainly investigated four doping sites in γ-graphyne: Fe atoms replacing single carbon atoms (C1 and C2 sites in Figure 1) and two adjacent carbon atoms (D1 and D2 sites in Figure 1). The structural stability of these doped systems (GY-Fe system), as well as the activation and dissociation behaviors of O_2 molecules on the substrates, were studied by using first-principles calculations. The results of ab initio molecular dynamics (AIMD) simulation, binding energies and electronic structural properties show that Fe atoms can be firmly anchored on the surface of γ-graphyne. In particular, O_2 molecules could be activated effectively on the surfaces of Fe-doped γ-graphyne and further dissociated with a low energy barrier according to the analysis of the optimal dissociation paths. Fe-doped γ-graphyne, as suggested by these results, might be a potential catalyst for oxidation reactions or the oxygen reduction reaction.

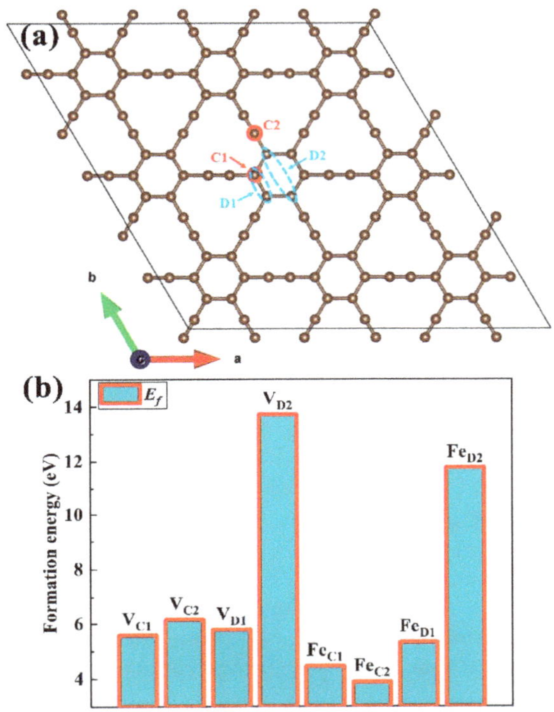

Figure 1. (**a**) Crystal structure of γ-graphyne and four substitution sites for Fe atom, which are labeled as C1, C2, D1 and D2, respectively. (**b**) Formation energies of defective γ-graphyne systems.

2. Computational Details

Density functional theory (DFT) was adopted as the theoretical basis for all first-principles calculations, and projector-augmented wave (PAW) pseudopotentials were employed too; these calculations were ultimately performed using the Vienna Ab initio Simulation Package (VASP) [32–37]. In this work, the generalized gradient approximation (GGA) served as the approach for addressing exchange–correlation interactions, and within this framework, the Perdew–Burke–Ernzerhof (PBE) functional was employed [38]. Two-dimensional γ-graphyne exhibits hexagonal symmetry, and the space group is P6/mmm. For the $3 \times 3 \times 1$ γ-graphyne super cell (containing 108 C atoms; see Figure 1), geometry optimization and electronic structure calculations employed Gamma-centered $2 \times 2 \times 1$ k-point meshes, with the plane-wave basis set assigned a kinetic energy cutoff of 450 eV. Geometry optimization employed a maximum Hellman–Feynman force of 0.05 eV/Å as its convergence criterion, with electronic iterations set to a convergence value of 1×10^{-5} eV. To simulate the adsorption and dissociation behavior of O_2 molecules on the substrate, a 15 Å thick vacuum layer was introduced to suppress interactions between adjacent layers. To describe the long-range van der Waals (vdW) interactions between the substrate and the adsorbate, the DFT-D3 dispersion-corrected method developed by Grimme was employed [39]. Adsorbed O_2 dissociation was studied using DFT calculations coupled with the climb image nudged elastic band (CI-NEB) method [40], with charge transfer and redistribution analyzed through Bader charge analysis [41] and the charge density difference (CDD) method.

For describing the interaction between oxygen molecules and the substrate, the adsorption energy of O_2 molecules designated as $E_{ad}(O_2)$ is defined by the following expression:

$$E_{ad}(O_2) = E(sub - O_2) - E(sub) - E(O_2) \quad (1)$$

Here, $E(sub\text{-}O_2)$ denotes the total energy of the substrate with adsorbed O_2, while $E(sub)$ and $E(O_2)$ represent the total energies of the bare substrate and the free O_2 molecule, respectively. The activation energy barrier (E_a) was introduced to describe both the minimum-energy reaction path (MEP) and the transition state (TS) during dissociation, with its definition given as

$$E_a = E_{TS} - E_{IS} \quad (2)$$

Here, $E(TS)$ denotes the total energy of the transition state (TS), while $E(IS)$ represents the total energy of the initial state (IS) in the reaction process, respectively.

3. Results and Discussion

3.1. Fe Atoms Doped into γ-Graphyne

In general, the formation process of the GY-Fe system should be described as the carbon atom being removed from the pristine γ-graphyne and then the impurity Fe atom being filled in carbon vacancy. As shown in Figure 1a, we selected four substitution sites for Fe atoms, including Fe replacing single carbon atoms of the benzene ring (the C1 site) and acetylene ring (the C2 site), respectively, and Fe substituting double carbon atoms of the benzene ring (D1 and D2 sites). According to some reports [42,43], in order to evaluate the process of impurity Fe replacing the C atom of pristine γ-graphyne, it is essential to calculate the formation energy of the GY-Fe system. The formation energy defined as

$$E_f = E_{GY-Fe} - E_{GY} - n_{Fe}\mu_{Fe} + n_C\mu_c \quad (3)$$

where E_{GY-Fe} and E_{GY} are the total energies of the GY-Fe system and pristine γ-graphyne. n_{Fe} (n_C) is the added (removed) number of Fe species (C), and μ_{Fe} (μ_c) is the chemical potential of the Fe (C) atom. μ_{Fe} is obtained from the Fe bulk material (face-centered cubic phase), and μ_c is obtained from pure γ-graphyne. For example, $\mu_{Fe} = \frac{E_{Fe}(bulk)}{4} = -8.16$ eV, $\mu_C = \frac{E_{GY}}{108} = -8.66$ eV. As a comparison, we also calculated the formation energy of the carbon vacancy in γ-graphyne (which will be discussed in detail; $E_f = E_{GY-Vacancy} - E_{GY} + n_C\mu_c$). The formation energies of defective γ-graphyne systems are displayed in Figure 1b. According to the results, we find that the values of formation energies are positive, which illustrates that the formation of these defective systems is an endothermic process. The order of the formation energy values of these defective systems is $E_f(Fe_{C2}) < E_f(Fe_{C1}) < E_f(Fe_{D1}) < E_f(V_{C1}) < E_f(V_{D1}) < E_f(V_{C2}) < E_f(Fe_{D2}) < E_f(V_{D2})$, which indicates that dopant Fe replacing carbon atoms at C2 and C1 sites is relatively easier than the formation of vacancy-like defects in pristine γ-graphyne. Besides that, the formation energy of Fe atoms occupying the D1 site is slightly less than that of the double-carbon vacancy (the D1 site), which indicates that Fe atom occupation at the D1 diatomic site by Fe atoms is relatively easy. However, the calculation results revealed that it is difficult to form D2-type double-carbon vacancy defects on the surface of γ-graphyne. Meanwhile, it is also difficult to form a defective surface with Fe atoms occupying the D2 sites. For comparison, we also investigated the adsorption and dissociation behaviors of oxygen molecules on the surface of GY-Fe (the D2 site).

Before investigating the activation and dissociation behaviors of O_2 on doped γ-graphyne surfaces with Fe atoms (GY-Fe), we first explored the structural stability and electronic properties of the GY-Fe systems. After structure relaxation, the stable doped substrates were obtained, as shown in Figure 2a–d. It is obvious that impurity Fe atoms could be bound by surrounding carbon atoms. The GY-Fe (C2 and D2 sites) systems face significant structural distortion, which is possibly because the interactions between Fe and neighboring carbon atoms are much stronger relative to the other two doped systems (C1 and D1 sites). The binding energies of these four doped systems are listed in Table 1; they are −6.32 eV, −7.47 eV, −5.66 eV and −7.14 eV for the C1 site, C2 site, D1 site and D2 site, respectively. In view of the distances between Fe and neighboring carbon atoms ($d_{Fe\text{-}C}$) being less than the sum of the atomic radii of Fe and C atoms, it can be considered that the formation of chemical bonds between Fe and adjacent carbon atoms occurs.

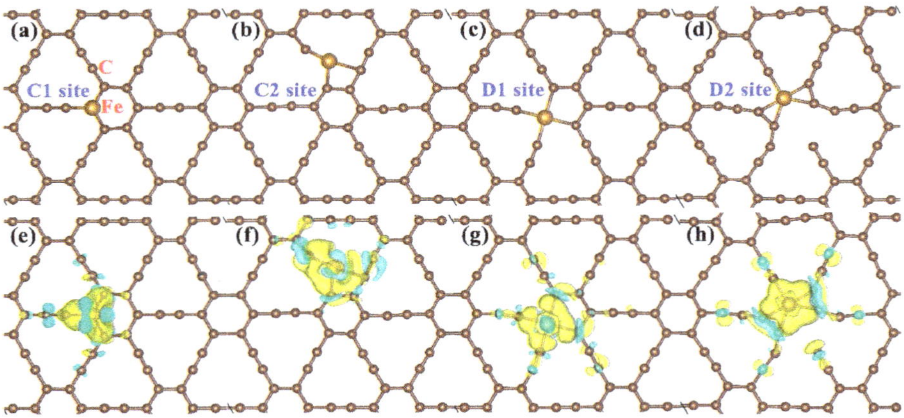

Figure 2. (**a–d**) Geometric configurations (top view) and (**e–h**) corresponding charge density difference (CDD) of GY-Fe systems. Saddle brown and ochre spheres represent C and Fe atoms, respectively. The yellow (blue) isosurface indicates the charge accumulation (depletion), and the isosurface value is taken to be 0.003 e/bohr^3.

Table 1. Calculated results for Fe-embedded graphyne systems, including the distance $d_{Fe\text{-}C}$ between Fe and neighboring carbon atoms, the binding energies of Fe atoms on graphyne (E_b), the charge transfer from graphyne to Fe atom (ΔQ).

Position	$d_{Fe\text{-}C}$ (Å)	E_b (eV)	ΔQ (e^-)
C1	1.80/1.75/1.80	−6.32	−0.78
C2	1.66/1.86/2.17	−7.47	−0.82
D1	1.84/1.84/1.98/1.98	−5.66	−1.11
D2	1.80/1.89/1.89/1.80/1.96	−7.14	−0.85

In order to gain insight into the interactions between Fe and the GY surface, the electronic properties of the four doped systems are discussed in detail. Figure 2e–h display the charge density difference (CDD) of the GY-Fe systems. Charge accumulations occur around the Fe-C chemical bonds, and it can be observed that a certain symmetric distribution of charges exists for the C1 site, D1 site, and D2 site doped systems. Furthermore, Bader charge analysis (Table 1) gives the amount (**Δ**Q) of charges transferring from GY to Fe atoms; they are about $-0.78\ e^-$, $-0.82\ e^-$, $-1.11\ e^-$ and $-0.85\ e^-$, respectively. Accompanied by charge transfers and charge redistributions, the formation of chemical bonding between Fe and neighboring carbon atoms takes place. In addition, as shown in Figure 3, it is observed that the Fe-3d orbital electronic states are strongly hybridized with the C-2p orbital electronic states, implying a strong interaction between them. Moreover, it is clear that some impurity peaks appear around the Fermi level due to the introduction of Fe into the GY surface, which mainly originates from Fe-3d orbital electronic states, implying that it may be more beneficial for an electron leaping into the conduction band from the valence band.

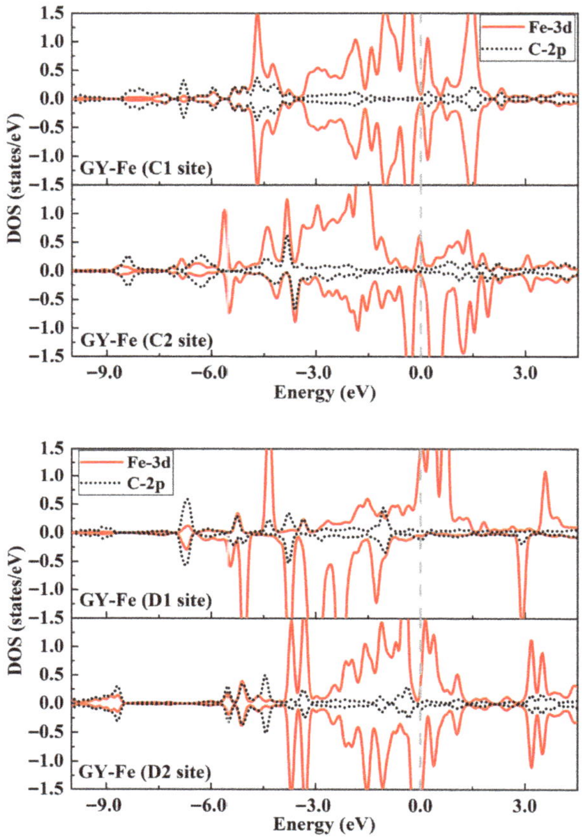

Figure 3. Partial density of states for impurity Fe and neighboring C atoms of GY-Fe systems (C1, C2, D1 and D2 sites). The Fermi level is set to zero.

Next, the dynamic stability of the GY-Fe systems was evaluated by ab initio molecular dynamics (AIMD) simulations with the NVT ensemble and the Nosé–Hoover thermostat [44], which is implemented in the DS-PAW calculation program under the Device Studio platform [35,45,46]. As shown in Figure 4, snapshots of the atomic structure and the total energy fluctuation within 10 ps of GY-Fe systems (the C1 site, C2 site, D1 site and D2 site) during the AIMD simulations at T = 300 K and 500 K (Figure S1) are presented. It is observed that the geometric structures of GY-Fe systems are well maintained. Moreover, the total energy shows slight oscillations at 300 K, with the oscillation range within ~2 eV. Thus, we think that the structural stability of GY-Fe systems can be maintained at room temperature. In summary, we believe that impurity Fe atoms could be firmly anchored in the γ-graphyne plane based on the analyses of binding energy, electronic properties, and dynamic stability simulations.

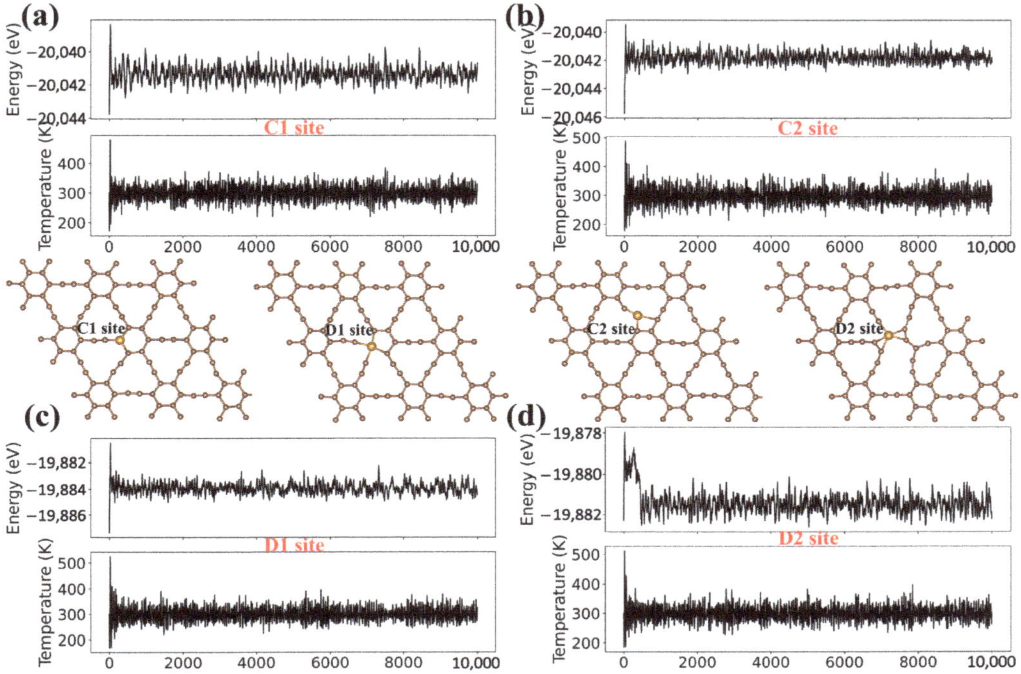

Figure 4. (**a–d**) Total energy fluctuation within 10 ps and snapshots of atomic structure of GY-Fe systems (C1 site, C2 site, D1 site and D2 site) during the AIMD simulations at T = 300 K.

3.2. Activation and Dissociation Behaviors of O_2 on GY-Fe Surfaces

The above calculation results indicate that there could be the formation of stable doped substrates by embedding Fe atoms in the γ-graphyne plane. Thereupon, we further explored the interactions between O_2 molecules and GY-Fe surfaces. The absolute value of the adsorption energy of O_2 molecules on the pristine γ-graphyne surface is below 0.007 eV, a result that suggests weak physisorption between O_2 and the surface. Notably, the adsorption capacity of GY surfaces for O_2 molecules could be enhanced remarkably when impurity Fe atoms are embedded in the plane. As listed in Table 2, the adsorption energies are −1.73 eV, −2.02 eV, −0.82 eV, and −2.21 eV for the C1 site, C2 site, D1 site, and D2 site, respectively, which implies that there should be chemical adsorption behavior of O_2 molecules on GY-Fe surfaces. Furthermore, when compared with the results of previous studies [47], it can be observed that the absolute value of the adsorption energy of O_2 molecules on the GY-Fe surface (C1 and C2 sites) obtained in the current study is relatively smaller. This indicates that the conventional DFT method tends to overestimate the adsorption behavior of O_2 molecules on the GY-Fe doped surfaces. In Figure 5a–d, the stable adsorption configurations of O_2 on GY-Fe surfaces are obtained after structural optimization. It can be clearly seen that O_2 molecules are fixed on the surfaces of GY-Fe systems through the bonding between Fe atoms and two oxygen adatoms. The lengths of Fe-O chemical bonds are about 1.83 Å for these four doped systems. Due to the chemical interactions between adsorbed O_2 and GY-Fe surfaces, the O-O bonds are nearly parallel to the substrates, and the bond lengths are elongated to 1.38 Å, 1.41 Å, 1.36 Å and 1.36 Å from 1.23 Å (gas phase) at the C1 site, C2 site, D1 site and D2 site, respectively. In fact, the lengths of O-O bonds are very close to the typical value of a superoxide (O_2^-) (the bond length is about 1.40 Å), which indicates that O_2 molecules are activated when they are adsorbed on the GY-Fe surfaces. Activated oxygen usually plays a key role in the oxidation reaction for toxic gas molecules [48] and the oxygen reduction reaction in hydrogen–oxygen fuel cells [24].

Table 2. Adsorption parameters of O_2 molecules on GY-Fe surfaces, including adsorption energy (E_a); the length of Fe-O chemical bond ($L_{Fe\text{-}O}$), the bond length of adsorbed O_2 molecules ($L_{O\text{-}O}$), charge transfer from the substrate to adsorbed O_2 molecules (ΔQ).

Position	E_a (eV)	$L_{Fe\text{-}O}$ (Å)	$L_{O\text{-}O}$ (Å)	ΔQ (e^-)
C_1	−1.73	1.90/1.83	1.38	0.61
C_2	−2.02	1.83/1.83	1.41	0.67
D_1	−0.82	1.86/1.86	1.36	0.52
D_2	−2.21	1.78/1.99	1.36	0.51

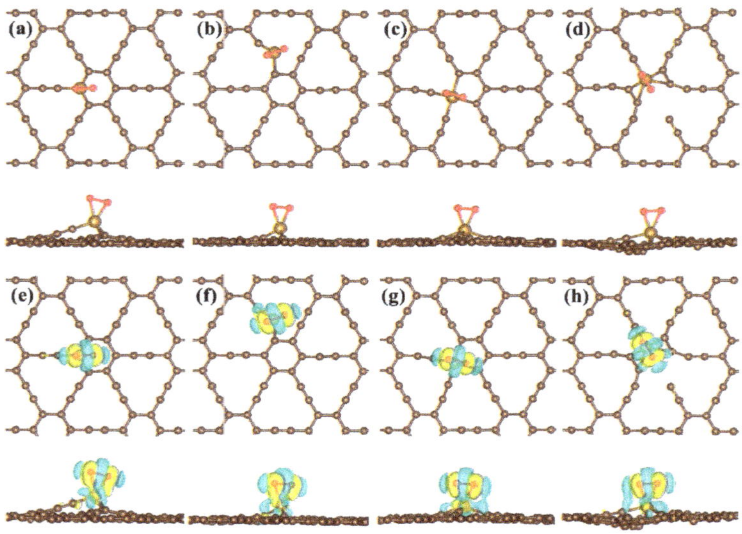

Figure 5. (**a**–**d**) Geometric configurations of and (**e**–**h**) charge density difference in O_2 molecule adsorption on the GY-Fe surfaces. Saddle brown, ochre and red spheres represent C, Fe and O atoms, respectively. The yellow (blue) isosurface indicates the charge accumulation (depletion), and the isosurface value is taken to be 0.005 e/bohr^3.

To further gain insight into the activation behaviors of O_2 molecules on GY-Fe surfaces, the spin-polarized partial density of states was calculated. As shown in Figure 6a,c, it is obvious that the electronic states ($O_{(1)}$ and $O_{(2)}$-2p orbitals) of adsorbed O_2 are sufficiently overlapping with the 3d orbitals of the dopant Fe atom, which directly confirms the formation of Fe-O chemical bonds between GY-Fe surfaces and the adsorbed O_2 molecules. Furthermore, the localization of electronic states (including $2\pi^*$, 1π and 5σ orbitals) is broken and energy levels are broadened in various degrees relative to free O_2 molecules (Figure 6b,d). Notably, the spin-down $2\pi^*$ orbital exhibits partial electron occupation, which implies that O_2 molecules act as electron acceptors during the adsorption process, and then the formation of superoxide anions on GY-Fe surfaces takes place. It can be confirmed from the Bader charge analysis (ΔQ), as listed in Table 2, that about 0.61 e^-, 0.67 e^-, 0.52 e^- and 0.51 e^- are transferred to the adsorbed O_2 from the substrates for the C1 site, C2 site, D1 site and D2 site, respectively. Besides that, charge redistributions occurred between adsorbed O_2 and Fe atoms because of the charges transferring between adsorbed O_2 and substrates. The results of the charge density difference (CDD) are shown in Figure 5e–h; charges accumulated around Fe-O chemical bonds, and charge depletion happened between the two O adatoms, which can serve as evidence that O_2 molecules are activated on GY-Fe surfaces. Moreover, according to the adsorption parameters of O_2 molecules on different doped surfaces (Table 2), the elongated lengths of O-O bonds and the amount of charge transfer exhibit a positive relationship. That is, the greater the amount of charge transferred to the adsorbed O_2, the larger the elongation of the O-O bond length. To provide a reasonable explanation, it is necessary to discuss the changes in the density of states of O_2 molecules on GY-Fe surfaces before and after adsorption. For example, as shown in Figure 6b (C2 site), it is obvious that the spin-down $2\pi^*$ orbital is partially filled with electrons (the peak of occupied states is located at -0.37 eV), but this is an unoccupied orbital for free O_2 molecules (the peak of unoccupied states is located at 1.83 eV). It is also because a certain amount of electrons were injected into the anti-bonding $2\pi^*$ orbital of adsorbed O_2 that the O-O bond length was significantly elongated. Similar features of electronic states for the other three doped systems (C1, D1 and D2 sites) also appeared. On the whole, the chemical inertness of the surface can be broken when impurity Fe atoms are doped in the γ-graphyne plane, and O_2 molecules are activated through charge transfer and charge redistribution between them and the substrates. However, there is still a certain amount of Fe-3d orbital electrons that need to be transferred to the carbon atoms in the substrates, and the amount of electron transfer is related to the Fe-C coordination structure (Table S1). In the C1 and C2 systems, O_2 molecules have a greater advantage in electron accepting ability compared to the substrate's carbon atoms, which means that the activation of O_2 molecules on doped surfaces is related to their surface structure.

Based on the above studies, the dissociation behaviors of adsorbed O_2 molecules on GY-Fe surfaces were further explored by the climbing image nudged elastic band (CI-NEB) method. As shown in Figures 7 and 8, the schematics of the minimum-energy reaction paths (MEPs) for O_2 dissociation on GY-Fe surfaces are described, including the dissociation energy barriers, as well as the configurations of the initial state (IS), transition state (TS) and final state (FS) in the reaction process. For the C1 site (Figure 7a IS-MS), adsorbed O_2 could be dissociated into two adsorbed O atoms with an energy barrier of about 0.97 eV: one O is located at the top site of the Fe atom, and the other O adatom moves to the vicinity of the para-position carbon atom. More electrons are transferred from Fe-3d orbitals to adsorbed oxygen molecules, occupying the $2\pi^*$ orbitals of adsorbed O_2 molecules, further elongating the O-O chemical bond until the bond gradually breaks. As shown in Figure 7b, it is very clear to observe that the density of state peak of the previously unoccupied spin-down O-2p ($O_{(1)}$-2p and $O_{(2)}$-2p) orbitals shifts towards the Fermi level and gradually becomes occupied. By comparing the total energy of the initial state (IS) with that of the middle state (MS), it can be concluded that the dissociation process is an endothermic reaction that absorbs about 0.50 eV of energy. Although the O-O bond has been broken at the saddle point, the two O adatoms remain bonded with the Fe atom in the dissociation process. As shown in Figure 7a (MS-FS), the adsorbed O atom located above the carbon atoms of the benzene ring can further migrate to the top site of the carbon atoms in the acetylene ring. During this process, only an energy barrier of 0.14 eV needs to be overcome, and a reaction heat of 0.57 eV is released. For the entire dissociation process (IS-MS-FS), the total energy change is calculated to be -0.07 eV, which indicates that the dissociation reaction of O_2 molecules on the surface of GY-Fe (the C1 site) can proceed (or is thermodynamically feasible).

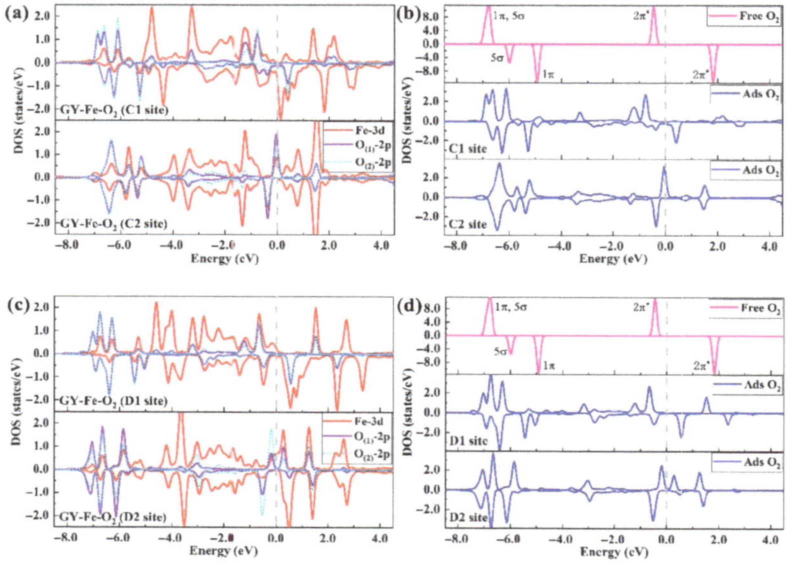

Figure 6. (**a**,**c**) The partial density of states for doped Fe-3d orbitals, as well as O-2p orbitals of adsorbed O_2 on the GY-Fe surface. The Fermi level is set to 0 eV. (**b**,**d**) The electronic density of states of free and adsorbed oxygen molecules.

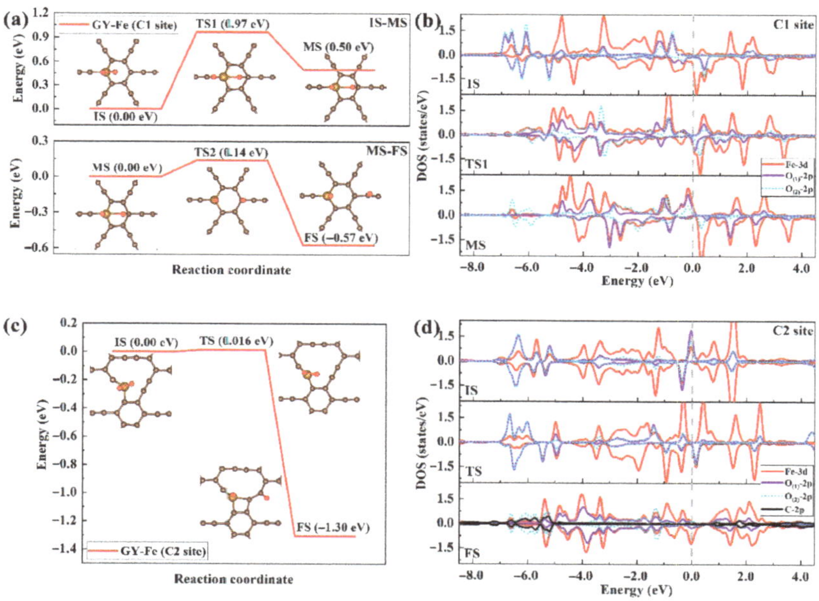

Figure 7. (**a**,**c**) The dissociation paths of adsorbed O_2 on GY-Fe surfaces (C1 and C2 sites), including the initial state (IS), transition state (TS), and final state (FS) configurations along the minimum-energy path (MEP). (**b**,**d**) The density of electronic states during the dissociation process. Saddle brown, ochre and red spheres represent C, Fe and O atoms, respectively.

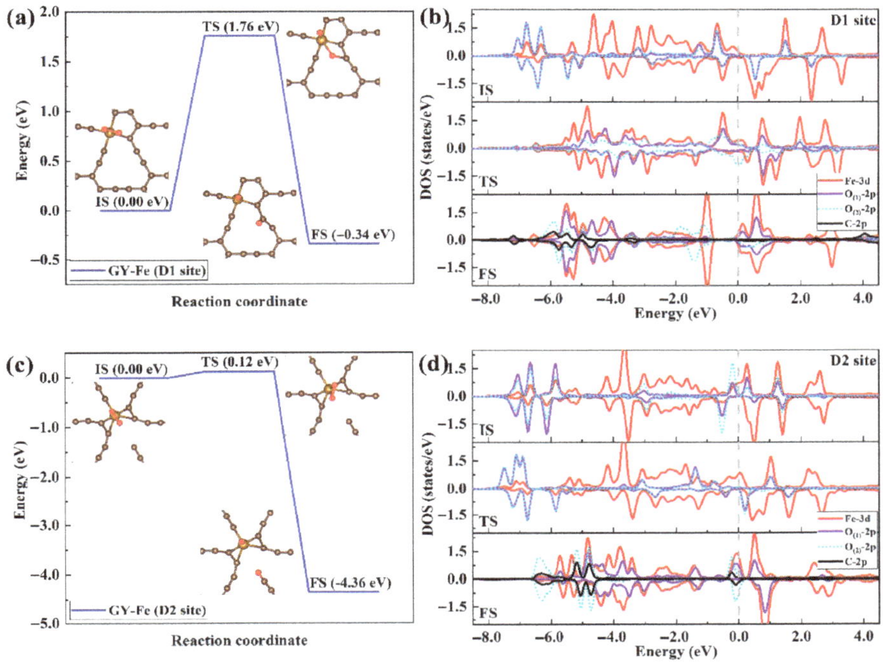

Figure 8. (**a**,**c**) The dissociation paths of adsorbed O_2 on GY-Fe surfaces (D1 and D2 sites), including the initial state (IS), transition state (TS), and final state (FS) configurations along the minimum-energy path (MEP). (**b**,**d**) The density of electronic states during the dissociation process. Saddle brown, ochre and red spheres represent C, Fe and O atoms, respectively.

As shown in Figure 7c, the O-O bond of adsorbed O_2 at the C2 site can be broken by overcoming an energy barrier of about 0.016 eV, and a more stable dissociation configuration (FS) is obtained through the release of about 1.30 eV. Analyzing the structure of FS, adsorbed O_2 on the GY-Fe surface (C2 site) completely dissociates, and one of the O atoms migrates to the vicinity of a carbon atom in the acetylene ring. Moreover, the reconstruction of chemical bonds between the impurity Fe and carbon atoms of the acetylene ring can also be observed. In addition, as explained regarding the relationship between the elongated lengths of O-O bonds and the amount of charge transfer in the previous section, we believe that the length of the O-O bond can be further elongated if more charges are transferred to the activated O_2. As shown in Figure 7d, it is clearly seen that there is a hybridization peak around at the Fermi level in the DOS of the C2 site, and the hybridization of electronic states (partially occupied spin-up electronic states) mainly originates from O-2p ($O_{(1)}$-2p and $O_{(2)}$-2p) orbitals (which are the main components of the $2\pi^*$ orbital (in Figure 6b C2 site)) of adsorbed O_2 and the Fe-3d orbital. The distribution of such electronic states is more favorable for the transfer of electrons from the GY-Fe substrate to the adsorbed O_2 molecules during the dissociation process. Subsequently, it can be observed that in the transition state, the peak of the electron density of states of the originally semi-occupied spin-up $2\pi^*$ orbitals gradually shifts beyond the Fermi level towards the valence band and is occupied by electrons. On the basis of these results, compared with the case of the C1 site, the dissociation reactions at the C2 site should be able to proceed spontaneously because of the lower energy barrier and greater amount of thermal energy released.

Similarly, we also simulated the dissociation process of O_2 molecules on the GY-Fe (D1 and D2 sites) systems. For the D1 site (Figure 8a), the energy barrier for adsorbed O_2 that separates into two O adatoms is about of 1.76 eV, and the total energy of system decreases by 0.34 eV after dissociation. The dissociation energy barrier is much higher than that of the above two systems, which indicates that the dissociation of O_2 molecules on the GY-Fe surface (at the D1 site) requires a large amount of external energy, although the dissociation process is an exothermic reaction. By analyzing the changes in the electronic density of states (DOSs) during the dissociation process (Figure 8b), we find that this process also involves charge transfer and the occupation of spin-down anti-bonding $2\pi^*$ orbitals and ultimately achieves the cleavage of the O-O chemical bond. However, the energy level difference between the hybridized overlapping peaks (resulting from the hybridization of spin-down O-2p and Fe-3d orbitals) located on both sides of the Fermi level is relatively large, which may be unfavorable for the charge transfer from Fe atoms to adsorbed O_2 molecules. In addition, in the GY-Fe system (the D1 site), the amount of charge transferred from Fe atoms to the substrate carbon atoms is higher than that in the other three systems (about 1.11 e^-). However, when oxygen molecules are adsorbed on its surface, the amount of charge transferred from the substrate to the oxygen molecules is the lowest (about 0.52 e^-). Therefore, it can be inferred that there is competition for charge transfer between the carbon atoms and the adsorbed O_2 molecules, which results in the lowest absolute value of the adsorption energy on its surface and the highest energy barrier during the dissociation reaction.

Different to the cases of the above three doped systems, for the D2 doped site (Figure 8c), the dissociation reaction of adsorbed O_2 molecules requires an extremely low energy barrier (0.12 eV), and the total energy decreases sharply by about 4.36 eV from the initial state (IS) to the final state (FS). The results illustrate that the dissociation of O_2 molecules on the GY-Fe (D2 site) surface can proceed continuously with the help of a large amount of released reaction heat. As shown in Figure 8d, it can be clearly seen that unoccupied electronic states of spin-up $O_{(1)}$-2p and $O_{(2)}$-2p orbitals (that is, the $2\pi^*$ orbital of adsorbed O_2 molecules in Figure 6c,d) and unoccupied states of the Fe-3d orbital completely overlap at the same level position (~0.27 eV). In fact, the electronic states of the Fe-3d orbital close to the Fermi level are partially occupied, and this energy region crosses the Fermi level, which indicates that more electrons can be further transferred into the unoccupied electronic states of the spin-up $2\pi^*$ orbital. Indeed, from the electronic density of state (DOS) results obtained in the dissociation process, the originally unoccupied 2p orbital DOS peaks were observed to gradually shift towards the Fermi level and become occupied. Therefore, based on the above results, we can conclude that the dissociation of O_2 molecules on the GY-Fe (D2 site) surface can be achieved with only a small amount of external energy.

4. Conclusions

In summary, the activation and dissociation behaviors of O_2 molecules on GY-Fe surfaces were systematically explored by using the first-principles calculations coupled with the CI-NEB method. The formation energy results show that the replacement of carbon atoms by Fe dopants (at C1, C2 and D1 sites) occurs relatively more easily than the formation of vacancy-like defects in γ-graphyne. From the results of electronic properties, the formation of chemical bonds between Fe atoms and adjacent carbon atoms indicates that Fe atoms are bound to the γ-graphyne surface. The results of ab initio molecular dynamics (AIMD) simulation confirmed that the GY-Fe systems (with C1, C2, D1, and D2 doped sites) can maintain structural stability at room temperature (300 K). O_2 molecules exhibit chemical adsorption behavior on the surfaces of all four doped substrates, while different doped surfaces show varying degrees of activation toward them. The results of electronic structure properties show that Fe-doped atoms transfer a certain amount of electrons to the adsorbed O_2 molecules, thereby causing an increase in the O-O bond length of the adsorbed O_2 molecules. More electrons are transferred from Fe-3d orbitals to adsorbed oxygen molecules, occupying the $2\pi^*$ orbitals of adsorbed O_2 molecules, further elongating the O-O chemical bond until the bond gradually breaks. Adsorbed O_2 molecules on the GY-Fe surfaces can be dissociated into two O adatoms with a lower energy barrier at the C1, C2, and D2 doped sites; the corresponding low dissociation energy barriers are about 0.97 eV, 0.016 eV, and 0.12 eV, respectively. Especially, these three dissociations are all exothermic reactions with the release of about 0.07 eV, 1.30 eV, and 4.36 eV, respectively. Therefore, the surfaces of γ-graphyne doped with Fe atoms are beneficial for the activation and dissociation of O_2 molecules, and this may serve as a feasible strategy for designing catalyst materials for oxidation reactions and oxygen reduction reactions.

Supplementary Materials: The following supporting information can be downloaded at: https://www.mdpi.com/article/10.3390/nano15191479/s1, Figure S1: (a–d) Total energy fluctuation within 10 ps and snapshots of atomic structure of GY-Fe systems (C1 site, C2 site, D1 site and D2 site) during the AIMD simulations at T = 500 K. Table S1: Charge transfer amounts between C-substrates, adsorption O_2 molecules and Fe atoms, as well as the bond lengths of adsorbed O_2 molecules ($L_{O\text{-}O}$).

Author Contributions: B.Z.: Conceptualization, Supervision, Funding Acquisition, Formal Analysis, Visualization, Writing—Review and Editing. J.Y., Z.X. and W.Y.: Data Curation, Investigation. P.G. and M.L.: Investigation, Resources, Software. H.Z.: Investigation. J.W.: Review and Editing, Resources, Software. All authors have read and agreed to the published version of the manuscript.

Funding: This work was support provided by the National Natural Science Foundation of China (NSFC) (Grant No. 11805295), the Key scientific research projects in higher education institutions in Henan Province (Grant No. 24A140028), the Funding scheme for young backbone teachers of Henan Province (Grant No. 2023GGJS109) and the Natural Science Foundation of Henan Province (Grants No. 232300421219).

Data Availability Statement: All data are included in the article and the Supplementary Materials.

Acknowledgments: We gratefully acknowledge HZWTECH for providing computation facilities. We gratefully acknowledge the National Supercomputing Center in Zhengzhou (NSCC-ZZ) for providing the computation resource.

Conflicts of Interest: The authors declare that they have no known competing financial interests or personal relationships that could have appeared to influence the work reported in this paper.

References

1. Kang, J.; Li, J.; Wu, F.; Li, S.-S.; Xia, J.-B. Elastic, electronic, and optical properties of two-dimensional graphyne sheet. *J. Phys. Chem. C* **2011**, *115*, 20466–20470. [CrossRef]
2. Li, Y.; Xu, L.; Liu, H.; Li, Y. Graphdiyne and graphyne: From theoretical predictions to practical construction. *Chem. Soc. Rev.* **2014**, *43*, 2572–2586. [CrossRef]
3. Kim, B.G.; Choi, H.J. Graphyne: Hexagonal network of carbon with versatile diraccones. *Phys. Rev. B* **2012**, *86*, 115435. [CrossRef]
4. Baughman, R.H.; Eckhardt, H.; Kertesz, M. Structure-property predictions for new planar forms of carbon: Layered phases containing sp^2 and sp atoms. *J. Phys. Chem. C* **1987**, *87*, 6687–6699. [CrossRef]
5. Kang, J.; Wei, Z.; Li, J. Graphyne and its family: Recent theoretical advances. *ACS Appl. Mater. Interfaces* **2018**, *11*, 2692–2706. [CrossRef]
6. Puigdollers, A.R.; Alonso, G.; Gamallo, P. First-principles study of structural, elastic and electronic properties of α-, β- and γ-graphyne. *Carbon* **2016**, *96*, 879–887. [CrossRef]

7. Zheng, E.; He, G.; Shang, C.; Chen, B.; Wang, Q.; Liu, Y. Insights into graphdiyne-supported single Ti for water dissociation reaction. *Comput. Theor. Chem.* **2022**, *1207*, 113499. [CrossRef]
8. Li, G.; Li, Y.; Liu, H.; Guo, Y.; Li, Y.; Zhu, D. Architecture of graphdiyne nanoscale films. *Chem. Commun.* **2010**, *46*, 3256–3258. [CrossRef] [PubMed]
9. Li, Y.; Li, Y. Chemical Modification and Functionalization of Graphdiyne. *Acta Phys.-Chim. Sin.* **2018**, *34*, 992–1013. [CrossRef]
10. Li, B.; Lai, C.; Zhang, M.; Zeng, G.; Liu, S.; Huang, D.; Qin, L.; Liu, X.; Yi, H.; Xu, F.; et al. Graphdiyne: A Rising Star of Electrocatalyst Support for Energy Conversion. *Adv. Energy Mater.* **2020**, *10*, 2000177. [CrossRef]
11. Li, Q.; Li, Y.; Chen, Y.; Wu, L.; Yang, C.; Cui, X. Synthesis of γ-graphyne by mechanochemistry and its electronic structure. *Carbon* **2018**, *136*, 248–254. [CrossRef]
12. Zuo, Z.; Li, Y. Emerging Electrochemical Energy Applications of Graphdiyne. *Joule* **2019**, *3*, 899–907. [CrossRef]
13. Akbari, F.; Vanani, A.R.; Darvishnejad, M.H. DFT study of the electronic and structural properties of single Al and N atoms and Al-N co-doped graphyne toward hydrogen storage. *Appl. Surf. Sci.* **2019**, *488*, 600–610. [CrossRef]
14. Liu, P.P.; Zhang, H.; Cheng, X.L.; Tang, Y.-J. External electric field: An effective way to prevent aggregation of Mg atoms on γ–graphyne for high hydrogen storage capacity. *Appl. Surf. Sci.* **2019**, *371*, 44–49. [CrossRef]
15. Gangan, A.; Chakraborty, B.; Ramaniah, L.M.; Banerjee, S. First principles study on hydrogen storage in yttrium doped graphyne: Role of acetylene linkage in enhancing hydrogen storage. *Int. J. Hydrogen Energy* **2019**, *44*, 16735–16744. [CrossRef]
16. Guo, Y.; Chen, Z.; Wu, W.; Liu, Y.; Zhou, Z. Adsorption of NO_x (x = 1, 2) gas molecule on pristine and B atom embedded γ-graphyne based on first-principles study. *Appl. Surf. Sci.* **2018**, *455*, 484–491. [CrossRef]
17. Mofidi, F.; Vanani, A.R. Sensing and elimination of the hazardous materials such as Sarin by metal functionalized γ-graphyne surface: A DFT study. *J. Mol. Liq.* **2019**, *286*, 110929. [CrossRef]
18. Darvishnejad, M.H.; Vanani, A.R. DFT-D3 calculations of the charge-modulated CO_2 capture of N/Sc-embedded graphyne: Compilation of some factors. *J. CO2 Util.* **2021**, *46*, 101469. [CrossRef]
19. Kang, B.; Shi, H.; Wu, S.; Zhao, W.; Ai, H.; Lee, J.Y. Revealing the importance of nitrogen doping site in enhancing the oxygen reduction reaction on β-graphyne. *Carbon* **2017**, *123*, 415–420. [CrossRef]
20. Gao, X.; Zhou, Y.; Liu, S.; Cheng, Z.; Tan, Y.; Shen, Z. Single cobalt atom anchored on N-doped graphyne for boosting the overall water splitting. *Appl. Surf. Sci.* **2020**, *502*, 144155. [CrossRef]
21. Kong, X.; Huang, Y.; Liu, Q. Two-dimensional boron-doped graphyne nanosheet: A new metal-free catalyst for oxygen evolution reaction. *Carbon* **2017**, *123*, 558–564. [CrossRef]
22. Fu, L.; Wang, R.; Zhao, C.; Huo, J.; He, C.; Kim, K.-H.; Zhang, W. Construction of Crembedded graphyne electrocatalyst for highly selective reduction of CO_2 to CH_4: A DFT study. *Chem. Eng. J.* **2021**, *414*, 128857. [CrossRef]
23. Gao, X.; Zhou, Y.; Cheng, Z.; Tan, Y.; Liu, S.; Shen, Z. Doping sp-hybridized B atoms in graphyne supported single cobalt atoms for hydrogen evolution electrocatalysis. *Int. J. Hydrogen Energy* **2019**, *458*, 781–789. [CrossRef]
24. Kattel, S.; Wang, G. Reaction Pathway for Oxygen Reduction on FeN_4 Embedded Graphene. *J. Phys. Chem. Lett.* **2014**, *5*, 452–456. [CrossRef]
25. Ma, D.; Tang, Y.; Yang, G.; Zeng, J.; He, C.; Lu, Z. CO catalytic oxidation on iron-embedded monolayer MoS_2. *Appl. Surf. Sci.* **2015**, *328*, 71–77. [CrossRef]
26. Yang, W.T.; Zhao, B.; Chen, W.; Guo, P.; Yin, L.; Li, M.; Zeng, H.; Zhang, M.; Wang, J. Theoretical investigation of adsorption and dissociation behaviors of H_2O molecules on the surfaces of Transition-metal doped γ-graphyne. *Mater. Today Commun.* **2023**, *37*, 107109. [CrossRef]
27. Deng, D.; Chen, X.; Yu, L.; Wu, X.; Liu, Q.; Liu, Y.; Yang, H.; Tian, H.; Hu, Y.; Du, P.; et al. A single iron site confined in a graphene matrix for the catalytic oxidation of benzene at room temperature. *Sci. Adv.* **2015**, *1*, e1500462. [CrossRef]
28. Zhang, N.; Zhou, T.; Chen, M.; Feng, H.; Yuan, R.; Zhong, C.; Yan, W.; Tian, Y.; Wu, X.; Chu, W.; et al. High-purity pyrrole-type FeN_4 sites as a superior oxygen reduction electrocatalyst. *Energy Environ. Sci.* **2020**, *13*, 111–118. [CrossRef]
29. Liu, S.; Cheng, L.; Li, K.; Yin, C.; Tang, H.; Wang, Y.; Wu, Z. RuN_4 Doped Graphene Oxide, a Highly Efficient Bifunctional Catalyst for Oxygen Reduction and CO_2 Reduction from Computational Study. *ACS Sustain. Chem. Eng.* **2019**, *7*, 8136–8144. [CrossRef]
30. Chen, X.; Qiao, Q.; An, L.; Xia, D. Why Do Boron and Nitrogen Doped α- and γ-Graphyne Exhibit Different Oxygen Reduction Mechanism? A First-Principles Study. *J. Phys. Chem. C* **2015**, *119*, 11493–11498. [CrossRef]
31. Jiao, D.; Zhang, D.; Wang, D.; Fan, J.; Ma, X.; Zhao, J.; Zheng, W.; Cui, X. Applying machine-learning screening of single transition metal atoms anchored on N-doped γ-graphyne for carbon monoxide electroreduction toward C1 products. *Nano Res.* **2023**, *16*, 11511–11520. [CrossRef]
32. Kresse, G.; Hafner, J. Ab initio molecular dynamics for liquid metals. *Phys. Rev. B* **1993**, *47*, 558. [CrossRef] [PubMed]
33. Kresse, G.; Hafner, J. Ab initio molecular-dynamics simulation of the liquid-metal-amorphous-semiconductor transition in germanium. *Phys. Rev. B* **1994**, *49*, 14251. [CrossRef]

34. Kresse, G.; Furthmüller, J. Efficient iterative schemes for ab initio total-energy calculations using a plane-wave basis set. *Phys. Rev. B* **1996**, *54*, 11169. [CrossRef]
35. Blöchl, P.E. Projector augmented-wave method. *Phys. Rev. B* **1994**, *50*, 17953–17979. [CrossRef]
36. Hohenberg, P.; Kohn, W. Inhomogeneous Electron Gas. *Phys. Rev.* **1964**, *136*, B864. [CrossRef]
37. Kohn, W.; Sham, L.J. Self-Consistent Equations Including Exchange and Correlation Effects. *Phys. Rev.* **1965**, *140*, A1133. [CrossRef]
38. Perdew, J.P.; Burke, K.; Ernzerhof, M. Generalized Gradient Approximation Made Simple. *Phys. Rev. Lett.* **1996**, *77*, 3865. [CrossRef]
39. Grimme, S.; Ehrlich, S.; Goerigk, L. Effect of the damping function in dispersion corrected density functional theory. *J. Comput. Chem.* **2011**, *32*, 1456–1465. [CrossRef]
40. Henkelman, G.; Uberuaga, B.P.; Jónsson, H. A climbing image nudged elastic band method for finding saddle points and minimum energy paths. *J. Chem. Phys.* **2000**, *113*, 9901. [CrossRef]
41. Henkelman, G.; Arnaldsson, A.; Jónsson, H. A fast and robust algorithm for Bader decomposition of charge density. *Comput. Mater. Sci.* **2006**, *36*, 354–360. [CrossRef]
42. Yun, J.; Zhang, Z.; Yan, J.; Zhao, W.; Xu, M. First-principles study of B or Al-doping effect on the structural, electronic structure and magnetic properties of γ-graphyne. *Comput. Mater. Sci.* **2015**, *108*, 147–152. [CrossRef]
43. Tang, Y.; Dai, X.; Yang, Z.; Liu, Z.; Pan, L.; Ma, D.; Lu, Z. Tuning the catalytic property of non-noble metallic impurities in grapheme. *Carbon* **2014**, *71*, 139–149. [CrossRef]
44. Nosé, S. A unified formulation of the constant temperature molecular dynamics Methods. *J. Chem. Phys.* **1984**, *81*, 511–519. [CrossRef]
45. Hongzhiwei Technology. *Device Studio, Version 2021A*; Hongzhiwei Technology: Shanghai, China, 2021. Available online: https://iresearch.net.cn/cloud-software (accessed on 15 September 2022).
46. Wang, M.; Wang, X.; Zheng, M.; Zhou, X. Improving Catalytic Activity of "Janus" MoSSe Based on Surface Interface Regulation. *Molecules* **2022**, *27*, 6038. [CrossRef] [PubMed]
47. Yang, W.T.; Zhao, B.; Li, C.Y.; Guo, P.; Li, M.; Ge, X.; Zhang, M.; Guan, X.; Wang, J. Enhanced gas molecules adsorption on γ-graphyne doped with Fe atom: A first-principles study. *Appl. Surf. Sci.* **2022**, *601*, 154083. [CrossRef]
48. Tang, Y.; Chen, W.; Chai, H.; Zhao, G.; Li, Y.; Ma, D.; Dai, X. Metal- and Nonmetal-Atom-Modified Graphene as Efficient Catalysts for CO Oxidation Reactions. *J. Phys. Chem. C* **2019**, *123*, 10926–10939. [CrossRef]

nanomaterials

MDPI

Article

A Study on the Microstructure and Mechanical Properties of Improved 25Ni-20Cr Steel via in Situ Testing

Penghui Lei [1], Xiaoyu Ji [1], Jiahao Chen [1], Yunhao Huang [2,*], Nan Lv [3,*], Yulin Fan [3], Yongchao Hou [3], Xinsheng Shi [3] and Di Yun [1]

[1] School of Nuclear Science and Technology, Xi'an Jiaotong University, Xi'an 710049, China; penghuilei@xjtu.edu.cn (P.L.); 2192112868@stu.xjtu.edu.cn (X.J.); chenjiahao101315@163.com (J.C.); diyun1979@xjtu.edu.cn (D.Y.)

[2] State Key Laboratory of Advanced Nuclear Energy Technology, Nuclear Power Institute of China, Chengdu 610041, China

[3] China National Heavy Machinery Research Institute Co., Ltd., Xi'an 710018, China; fylyhx@sina.com (Y.F.); yongchaohou-001@163.com (Y.H.); shizmw123@foxmail.com (X.S.)

* Correspondence: huangyunhao@mail.nwpu.edu.cn (Y.H.); cpzremix@163.com (N.L.)

Abstract: To meet the application requirements for structural components in Gen-IV nuclear reactors, it is essential to improve the high-temperature mechanical properties of 25Ni-20Cr (S35140) austenitic stainless steel. In this research, an improved austenitic stainless steel (N-S35140), derived from S35140 steel, was investigated. The scanning transmission electron microscopy (STEM) results indicate that the addition of titanium (Ti) microalloying elements to S35140 steel led to the precipitation of new strengthening nano phases, including M(C, N), MC, MN and Ti(C, N), in N-S35140. These precipitates effectively compensated for the loss of high-temperature strength resulting from the substantial reduction in carbon content. During the in situ transmission electron microscopy (TEM) compressive process at room temperature, the yield strength of N-S35140 steel is 618.4 MPa. At room temperature, the tensile strength of N-S35140 steel is 638.5 MPa, with a yield strength of 392.8 MPa and an elongation of 32.7%, which surpasses those of S35140 steel at room temperature. N-S35140 steel exhibits a tensile strength of 330.6 MPa, a yield strength of 228.2 MPa, and an elongation of 51.4% during the in situ scanning electron microscopy (SEM) tensile test conducted at 650 °C. As a consequence, the improved N-S35140 steel demonstrates significantly enhanced mechanical properties compared to the original S35140 steel, positioning it as a promising candidate for structural components in Gen-IV nuclear reactors.

Keywords: Gen-IV nuclear reactors; N-S35140 austenitic steel; mechanical properties; TEM in situ compressive test; SEM in situ tensile test

1. Introduction

S35140 (based on Fe-25Ni-20Cr Steel/Alloy 709) austenitic steel is considered as a promising candidate structural material for the components of Gen-IV sodium-cooled fast neutron reactor (SFR) and supercritical water reactor (SCWR), due to its exceptional mechanical properties [1–4]. The high strength and stability of S35140 steel result from its higher carbon (C) and nickel (Ni) content design, with the primary strengthening phases being (Nb,Ti)CN and M23C6 [5]. Kruizenga et al. [6] reported that S35140 steel exhibited excellent corrosion resistance. However, in Generation IV nuclear reactors, where the operating temperature typically reaches 550 °C or higher [7], the higher carbon content in the S35140 steel negatively impacts its high-temperature aging stability and long-term corrosion resistance. At higher operating temperatures, carbides such as M23C6, M6C tend to precipitate within the matrix and along grain boundaries [8]. The M23C6 phase, in particular, undergoes significant coarsening above 600 °C, which adversely affects high-temperature strength and resistance to supercritical water corrosion [9–11]. Meanwhile, in a high-temperature steam environment, chromium (Cr) and oxygen (O) are prone to form volatile chromium hydrides with water vapor, further limiting the working temperature and service environment of austenitic stainless steel [12,13]. Therefore, it is essential to improve and enhance the properties of S35140 austenitic stainless steel to better meet the demanding requirements of high-temperature and corrosive environments in nuclear applications.

The properties of stainless steel in high-temperature service are affected by the content of alloying elements in the material [14–17]. Compared to the Cr_2O_3 film, the Al_2O_3 film formed by adding aluminum (Al) to austenitic stainless-steel exhibits superior thermal stability and enhanced corrosive resistance at high temperatures [18–21]. The addition of elements such as titanium (Ti) and niobium (Nb) improves the oxidation resistance, the corrosion resistance, and the mechanical properties of stainless steel [22–26]. Thus, an improved austenitic stainless steel (N-S35140) was developed based on S35140 austenitic steel using a microalloying design method, with the aim of meeting the application requirements for structural components in Gen-IV nuclear reactors [27]. This microalloying method involved a significant reduction in C content, along with the regulation of microalloying elements such as N and Nb, and the incorporation of controlled amounts of additional elements like Al and Ti. This approach compensates for the loss of high-temperature strength resulting from the significant reduction in C content by promoting the formation of new strengthening phases. Further research is required to comprehensively investigate the microstructure and mechanical properties of N-S35140 steel.

In this paper, the microstructural morphologies and elemental distributions of the N-S35140 are investigated using TEM equipped with Energy-Dispersive X-ray Spectroscopy (EDS) systems [28]. SEM in situ tensile tests and TEM in situ compressive tests were performed to systematically investigate the evolution of the microstructure and mechanical properties of N-S35140 at various service temperatures. Electron backscatter diffraction (EBSD) was used to analysis changes in grain orientation.

2. Materials and Methods

The chemical composition of N-S35140 steel is presented in Table 1. In this research, all experimental samples were melted at 1600 °C using a vacuum induction furnace under a vacuum of less than 1×10^{-4} Pa. The resulting ingots were forged at a holding temperature of 1200 °C for 2 h, with a final forging temperature of 1100 °C and a forging ratio of 3:1. Following forging, the ingots were held at 1200 °C for 2 h and subsequently rolled with a deformation of 65%. The final rolling temperature was approximately 1050 °C. After rolling, the forgings were quenched by water cooling.

Table 1. Actual chemical composition of experimental steel (wt. %) [27].

Experimental Steel	**Cr**	**Ni**	**Nb**	**Al**	**Si**	**C**
S35140	20.0~22.0	25.0~27.0	0.25~0.75	-	0.75	0.1
N-S35140	19.59	24.2	0.77	0.02	0.40	0.03
Experimental Steel	Mo	N	Ti	Mn	Fe	
S35140	1.0~2.0	0.08~0.20	-	1.0~3.0	Bal.	
N-S35140	1.69	0.1	0.04	-	Bal.	

Grain orientation was analyzed using SEM with EBSD under a vacuum of 5×10^{-5} Pa (Gemini 460, Carl Zeiss AG, Oberkochen, Germany). The electrochemical double-spray method was employed to prepare samples for TEM microstructure characterization under a vacuum of 1.1×10^{-7} Pa (Talos F200X, Thermo Fisher Scientific, Waltham, MA, USA). The preparation of samples for TEM in situ compressive tests was carried out using a focused ion beam (FIB) under a vacuum of 4.9×10^{-5} Pa (Thermo Scientific, Helios 5 UX, Thermo Fisher Scientific, Waltham, MA, USA). The dimensions of the TEM in situ compressive column sample are φ 200 nm × 800 nm. In situ compression experiments were conducted using a PI-95 in situ mechanics sample holder from Bruker. The sample holder was mechanically controlled, piezoceramic, and sensor-controlled in multiple dimensions to ensure precise nanoscale movement of the probes. The probe indentation was controlled by displacement, with a maximum compression displacement of 300 nm and a displacement rate of 10 nm/s. The loading sequence followed a function of 30 s for loading, 10 s for retention, and 30 s for unloading. Room-temperature and high-temperature in situ tensile tests of N-S35140 steel were carried out using an SEM in situ tensile platform. The tensile test temperatures were 25 °C (room temperature), 350 °C, and 650 °C. The heating time to the target temperature was 2–3 h, followed by a 1 h holding period. The dimensions of the in situ room temperature tensile samples are shown in Figure 1, while the dimensions of the in situ high-temperature tensile samples are illustrated in Figure 2. Wire-cutting and electropolishing methods were used to prepare the samples for SEM in situ testing. The carrier platform used for the SEM in situ tensile tests was mechanically controlled in a constant step mode for sample stretching. The room-temperature displacement speed was set at 100 μm/min. For high-temperature tests at 350 °C and 650 °C, the displacement speeds were 60 μm/min and 30 μm/min, respectively. The yield strength, tensile strength, and elongation of the tensile specimens were determined using an extensometer.

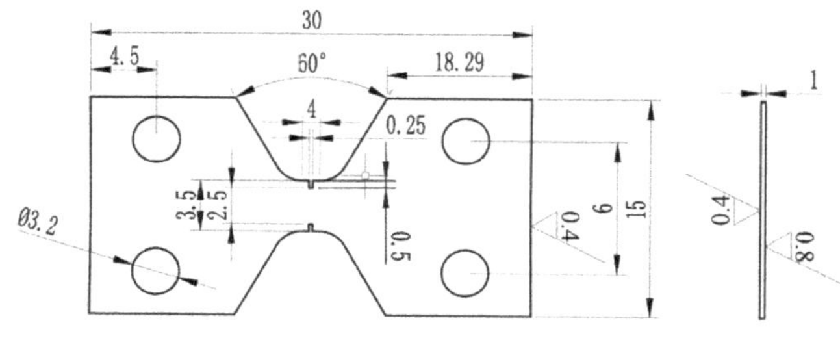

Figure 1. Dimensions of room-temperature tensile specimen (unit: mm).

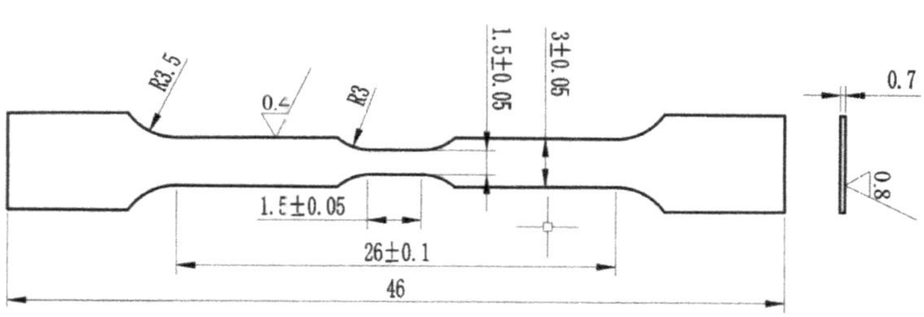

Figure 2. Dimensions of high-temperature tensile specimen (unit: mm).

3. Results and Discussion

3.1. Microstructures Characterization

The microstructure morphologies of N-S35140 steel are presented in Figure 3. Numerous nanoscale precipitates with distinct interfaces to the matrix grains are visible within the matrix, as shown in Figure 3a. A significant number of precipitates are observed both within the grains and along the grain boundaries, while twin boundaries containing parallel dislocations are evident within the crystals in Figure 3b. The precipitates with different morphologies are observed in Figure 3c. Selected-area electron diffraction (SAED) of grain A, shown in Figure 3d, reveals that the matrix grain exhibits a face-centered cubic (FCC) crystal structure, corresponding to the [0 1 1] band-axis diffraction pattern. The diffraction spots closest to the central transmission spot are attributed to the (−2 0 0) and (−1 1 −1) crystal planes.

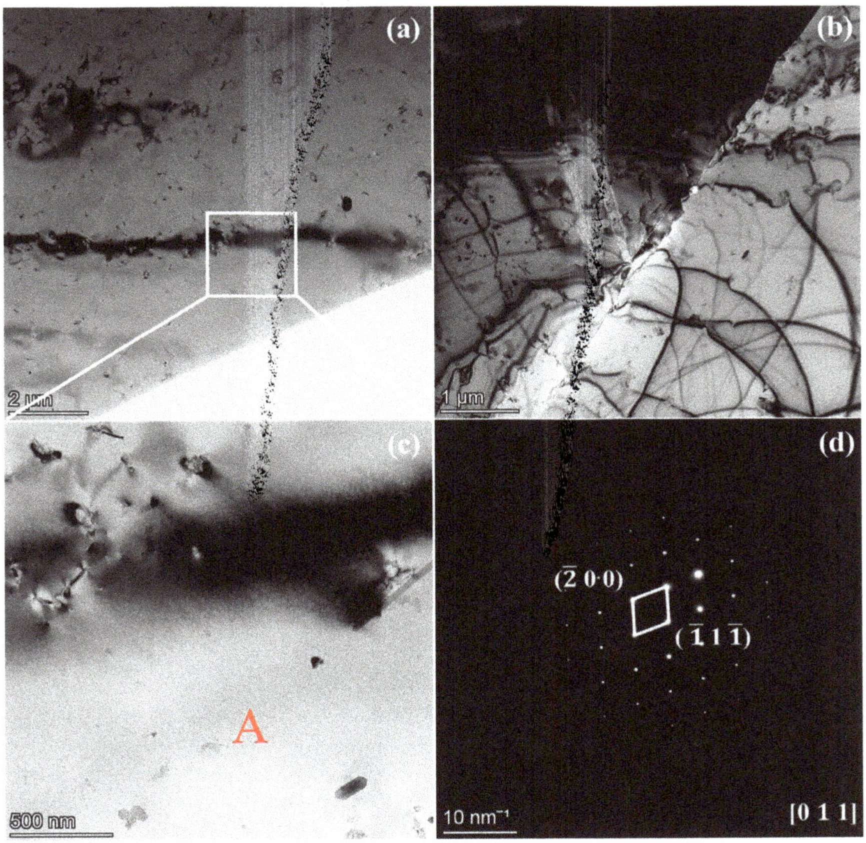

Figure 3. The microstructure morphologies of N-S35140 steel (**a**–**c**); (**d**) SAED of grain A.

Figure 4 presents the STEM image and EDS mapping results of the intragranular structure in N-S35140 steel. The EDS analysis reveals that precipitate P1 is enriched with elements such as Fe, Cr, Mo, Ni, C, N, and Ti, while precipitate P2 shows elemental enrichment of Cr, Mo, Nb, C, N, and Ti. Both of these precipitates are identified as M(C, N) carbonitrides. The remaining precipitated phases (indicated by blue circle), with sizes ranging from approximately 50 to 100 nm, are enriched in Mo, Nb, and C elements, and are identified as MC-type carbide precipitates. MC-type carbide precipitates are commonly employed to enhance creep resistance at elevated temperatures in advanced heat-resistant austenitic stainless steels [29].

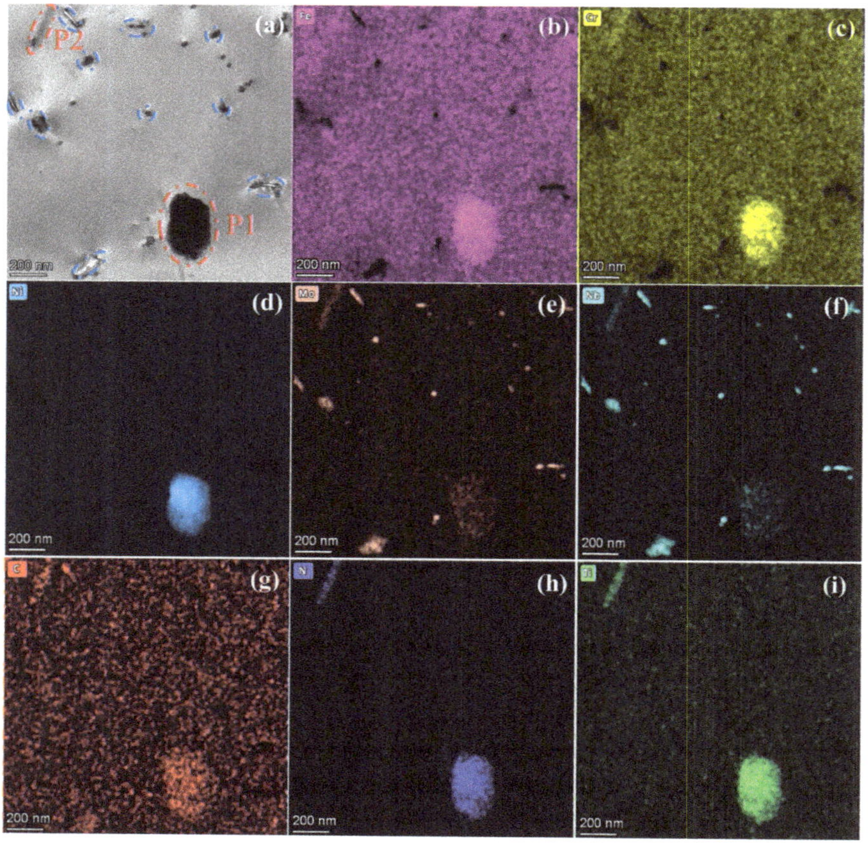

Figure 4. STEM image (**a**) and EDS results (**b–i**) of the intragranular structure in N-S35140 steel.

The EDS analysis images of S35140 steel near the grain boundary are shown in Figure 5. Figure 5a presents the STEM bright-field image, while Figure 5b–i display the EDS mapping results. The observation indicates that the precipitate P3 is enriched in Cr, N, and Ti elements, corresponding to an MN nitride precipitation phase. The elongated precipitate P4, enriched in C, N, and Ti elements, is identified as the Ti(C, N) carbonitride. The remaining precipitates (indicated by yellow circle), with sizes ranging from 50 to 100 nm, are enriched with Mo, Nb, and C elements, and are identified as MC carbide precipitates. The beneficial trace elements that segregate at the grain boundaries can enhance the grain-boundary bonding by altering the interatomic bonding state [30] and improve strength or hardness by suppressing grain-boundary migration and blocking the dislocation motion [31,32], ultimately leading to enhanced high-temperature strength of the alloy. Therefore, by adding Ti microalloying element, new strengthening phases such as MN and Ti(C, N) were precipitated in N-S35140 steel, effectively compensating for the loss of high-temperature strength resulting from the significant reduction in carbon content. Precipitates at grain boundaries hinder grain-boundary migration, thereby delaying matrix-softening processes, which positively contributes to improved creep resistance [33].

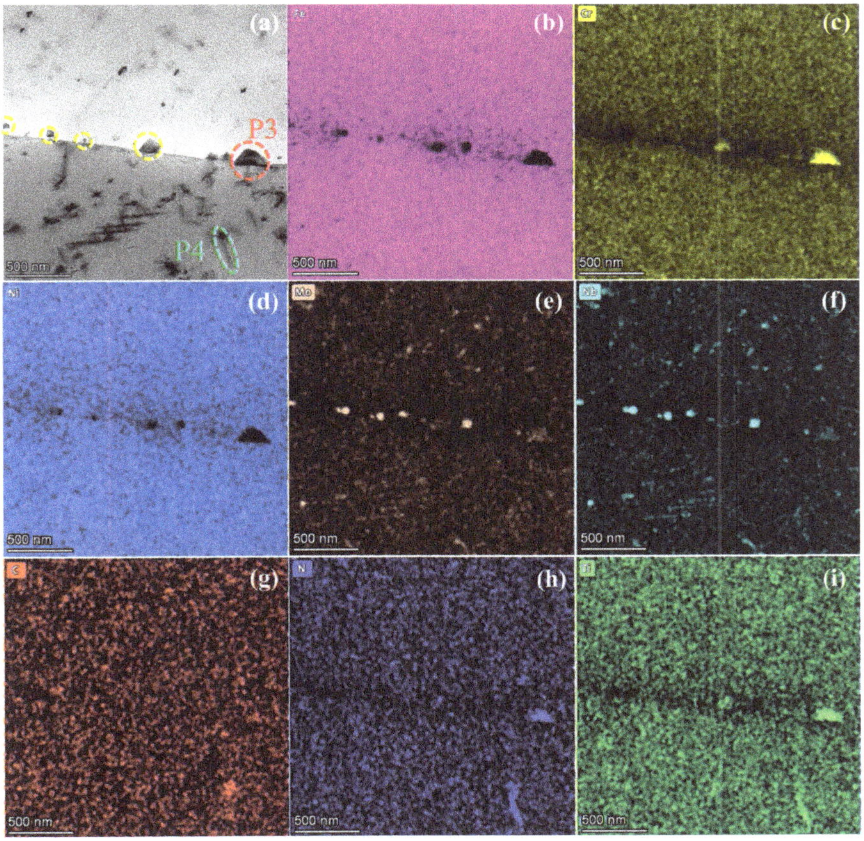

Figure 5. STEM image (**a**) and EDS results (**b–i**) of N-S35140 steel near the grain boundary.

3.2. Compression Properties

The TEM in situ compression process of N-S35140 steel at room temperature is shown in Figure 6. Specifically, Figure 6a–c presents the load–displacement curves and TEM images corresponding to the starting loading, compression, and unloading stages, respectively. At an indenter displacement of 48 nm, the indenter fully contacts the nano pillar, with a maximum load of 43 μN. The plastic flow of the N-S35140 steel nano pillar remained stable throughout the deformation process, with uniform deformation. No significant abrupt changes in displacement were observed.

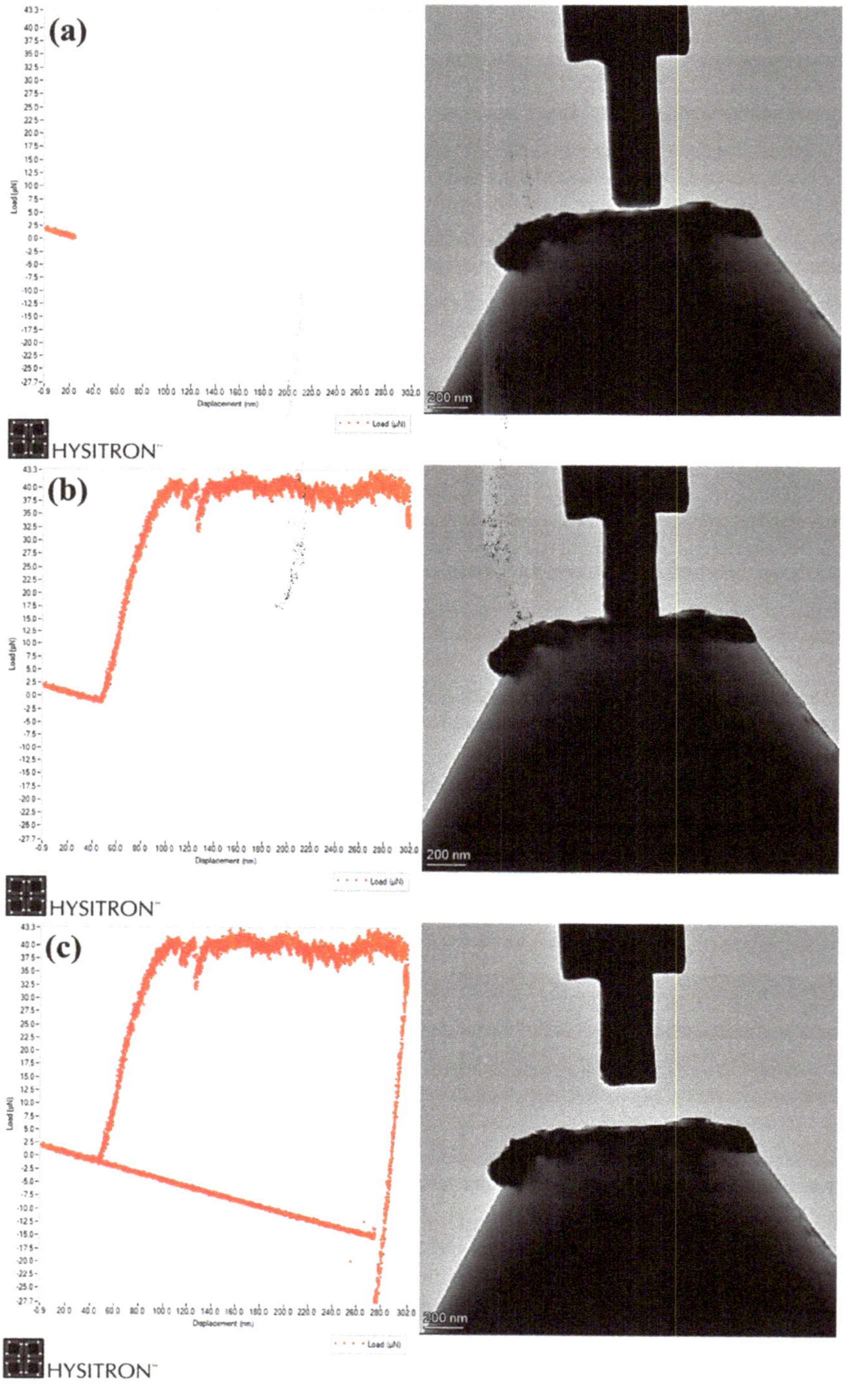

Figure 6. TEM in situ compression process of N-S35140 steel nano pillar: (**a**) starting loading stage; (**b**) compression process stage; (**c**) unloading stage.

The stress–strain curves of N-S35140 steel during TEM in situ compression at room temperature is displayed in Figure 7. The compressive yield strength is determined to be 618.4 MPa, based on the intersection of the tangent lines of the elastic and plastic deformation regions, which is significantly higher than the compressive yield strength of AISI 304L stainless steel [34]. Low yield strength is often a significant limitation of austenitic stainless steel. However, N-S35140 steel, with its higher compressive yield strength, demonstrates superior shape and dimensional stability. It is worth noting that in the case of TEM in situ testing, the size effect (micro- and nanoscale) is primarily attributed to the reduction in sample size, which leads to a decrease in the number of dislocation sources. This shortage of dislocation sources makes it more difficult to activate dislocation slip for plastic deformation, requiring larger loads to achieve plastic deformation. As a result, the yield strength of N-S35140 steel during TEM in situ compression is higher than that of AISI 304L with classical size due to the size effect [35].

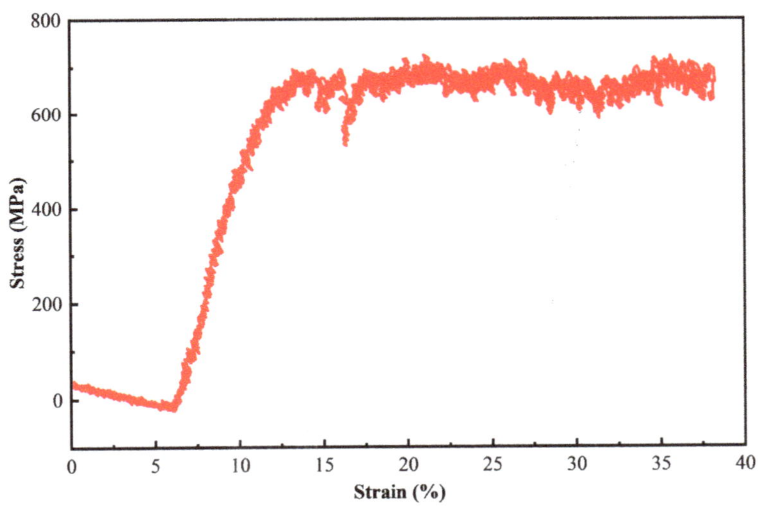

Figure 7. The stress–strain curves of N-S35140 steel during TEM in situ compression at room temperature.

3.3. Tensile Properties and Microstructure

The stress–strain curves and strength of N-S35140 steel obtained during SEM in situ tensile testing at different temperatures are presented in Figure 8. The size of N-S35140 steel samples for SEM in situ tensile tests is on the millimeter scale, making it suitable for comparison with classical tensile tests. At room temperature, the tensile strength of N-S35140 steel is 638.5 MPa, with a yield strength of 392.8 MPa and an elongation of 32.7%. According to ASTM (A240/A240M-19) [36], the tensile strength, yield strength, and elongation of S35140 steel at room temperature are 620 MPa, 275 MPa, and 30%, respectively. It is evident that the tensile strength, yield strength, and elongation of N-S35140 steel surpass those of S35140 steel at room temperature. During SEM in situ tensile testing at 350 °C, N-S35140 steel exhibits a tensile strength of 481.7 MPa, a yield strength of 234.9 MPa, and an elongation of 66.5%. When the testing temperature is increased to 650 °C, the tensile strength decreases to 330.6 MPa, with a yield strength of 228.2 MPa and an elongation of 51.4%. The yield strength and elongation of N-S35140 steel are superior to those of Alloy 709 steel and AISI 316L at elevated temperatures [37,38], highlighting its enhanced shape and dimensional stability, as well as improved ductility and toughness under load.

The segregation of interstitial atoms, such as carbon and nitrogen, in the tensile-stress zone creates a pinning effect on dislocations [39]. As the temperature increases, atomic thermal vibrations intensify, aiding in overcoming this resistance [40]. Consequently, the strength of N-S35140 steel decreases with rising temperature. The serrated curve observed on the stress–strain curve indicates the presence of dynamic strain aging (DSA) at high temperatures.

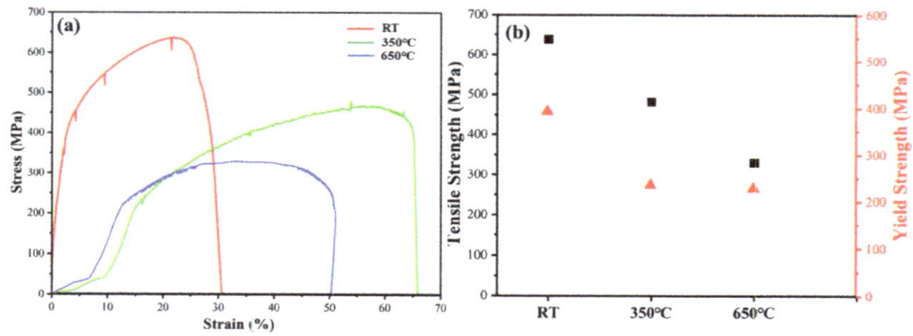

Figure 8. The stress–strain curves (**a**) and tensile strength and yield strength (**b**) of N-S35140 steel during SEM in situ tensile testing at different temperatures.

The EBSD result, based on inverse pole figure (IPF) map along the Z-axis of N-S35140 steel prior to SEM in situ tensile testing at 350 °C, is shown in Figure 9a. The average grain size of N-S35140 steel is approximately 200 μm. The matrix of N-S35140 steel exhibits an austenitic structure, predominantly consisting of equiaxial crystals before tensile testing. The overall observation reveals a uniform grain distribution and the presence of a twin-crystal structure within some grains, which is primarily attributed to the rolling process of N-S35140 steel. After tensile fracture, the IPF map in Figure 9b shows that the grains have deformed along the tensile direction, with twins still present. Significant grain deformation along the tensile direction is further evidence, shown in Figure 9d, for N-S35140 steel after tensile fracture at 650 °C. Significant grain elongation is observed after deformation, accompanied by an increase in the proportion of grains exhibiting (1 0 1) crystal orientation. Austenitic stainless steels generally have relatively low stacking-fault energy [41], which makes them prone to the formation of stacking faults and twins during tensile deformation. The number of twins decreases significantly after deformation at different temperatures, suggesting that twin-induced plasticity (TWIP) is not the sole contributor to the material's ductility.

The fracture surfaces of N-S35140 steel after SEM in situ tensile testing at room temperature are shown in Figure 10a. Numerous small dimples are observed, indicating a typical ductile fracture mechanism. As the testing temperature increases to 350 °C (Figure 10b) and 650 °C (Figure 10c), the dimples become larger, deeper, and less densely distributed, reflecting the influence of elevated temperatures on the fracture morphology (Table 2). The density, size, and depth of dimples are directly related to the plasticity of the material fracture [42]. Materials with poor plasticity tend to form smaller and shallower ductile dimples during fracture, whereas materials with good plasticity produce larger and deeper dimples. Therefore, the observations suggest that N-S35140 steel exhibits better plasticity at high temperatures compared to room temperature, which aligns with the results shown in Figures 8 and 9.

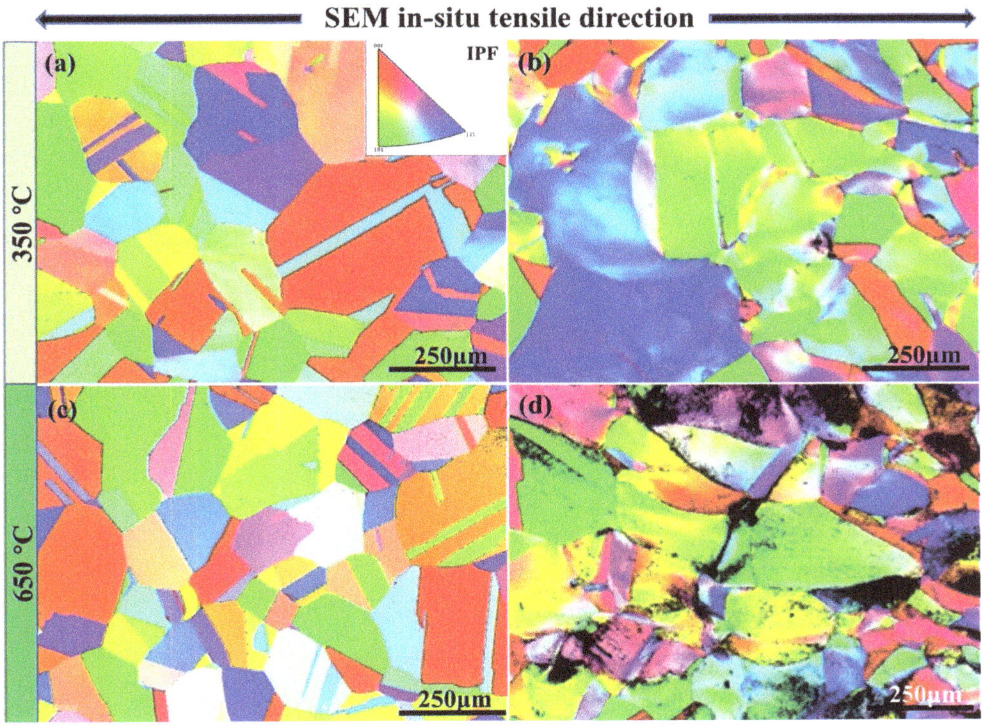

Figure 9. The EBSD results based on inverse pole figure (IPF) map along the *Z*-axis of N-S35140 steel: initial morphologies (**a**,**c**); SEM in situ tensile testing at 350 °C (**b**) and at 650 °C (**d**).

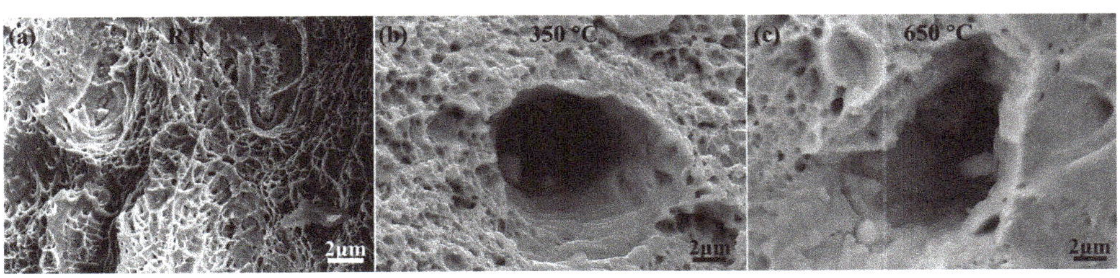

Figure 10. Morphologies of N-S35140 steel during in situ tensile fracture at (**a**) room temperature; (**b**) 350 °C; and (**c**) 650 °C.

Table 2. Summary of testing conditions, areal density of dimples, and average dimple diameter of N-S35140 steel.

Testing Temp. (°C)	Average Dimple Diameter	Areal Density of Dimple ($\times 10^{11}/m^2$)
TR	540 nm	12.2
350	695 nm	4.05
650	899 nm	3.15

4. Conclusions

In this research, the microstructure of the improved N-S35140 steel for nuclear reactors was observed using a microalloying design method, then analyzed and characterized, while the mechanical properties were researched and analyzed. The following main conclusions were obtained:

(1) By the regulation of microalloying elements in S35140 steel, new strengthening phases M(C, N), MC, MN and Ti(C, N) were precipitated, which compensated for the loss of high-temperature strength after a significant reduction in carbon content. Typically, there is a positive effect of the precipitates distributed along the grain boundaries on the mechanical properties of NS steels. Improved N-S35140 steel has numerous precipitated phases distributed both within the grains and at the grain boundaries. In addition, planar twinning dislocations were observed inside the crystals.
(2) The compressive yield strength of N-S35140 steel during TEM in situ compression at room temperature is 618.4 MPa, which is significantly higher than the compressive yield strength of AISI 304L stainless steel. N-S35140 steel, with its higher compressive yield strength, demonstrates superior shape and dimensional stability compared to AISI 304L stainless steel under load. At room temperature, the tensile strength of N-S35140 steel is 638.5 MPa, with a yield strength of 392.8 MPa and an elongation of 32.7%, which surpasses those of S35140 steel at room temperature. During SEM in situ tensile testing at 350 °C, N-S35140 steel exhibits a tensile strength of 481.7 MPa, a yield strength of 234.9 MPa, and an elongation of 66.5%. When the testing temperature is increased to 650 °C, the tensile strength decreases to 330.6 MPa, with a yield strength of 228.2 MPa and an elongation of 51.4%. The yield strength and elongation of N-S35140 steel are superior to those of Alloy 709 steel and AISI 316L at elevated temperatures.
(3) The grains from the IPF map have deformed along the tensile direction, with twins still present after tensile fracture at 350 °C. Significant grain deformation along the tensile direction is further evident after tensile fracture at 650 °C. Numerous small dimples are observed after SEM in situ tensile testing at room temperature, indicating a typical ductile fracture mechanism. As the testing temperature increases to 350 °C (Figure 10b) and 650 °C (Figure 10c), the dimples become larger, deeper, and less densely distributed. N-S35140 steel exhibits better plasticity at high temperatures compared to room temperature.

In general, the results of this study indicate that N-S35140 steel, as a potential structural material for Gen-IV nuclear reactors, exhibits excellent tensile strength, yield strength, and elongation at high temperatures. Long-term aging, corrosion testing, irradiation, and creep properties of N-S35140 steel will be investigated in further work.

Author Contributions: Writing—original draft preparation, P.L.; Methodology, X.J.; Conceptualization, J.C.; Writing—review and editing, Y.H. (Yunhao Huang); Writing—review and editing, N.L.; Data curation, Y.F.; Methodology, Y.H. (Yongchao Hou); Conceptualization, X.S.; Investigation, D.Y.; All authors have read and agreed to the published version of the manuscript.

Funding: This work is supported by the National Natural Science Foundation of China (12205232), Scientific research program for young talent of China National nuclear corporations (NPIC-k301007038), Research Project of China Machinery Industry Group Co., Ltd. (Grant No. ZDZX2024-28) and Research Project of China Machinery Industry Group Co., Ltd. (Grant No. ZDZX2024-12).

Data Availability Statement: The data presented in this study are available on request from the corresponding authors.

Acknowledgments: We thank Zijun Ren at Instrument Analysis Center of Xi'an Jiaotong University for their assistance with SEM and image analysis.

Conflicts of Interest: Authors Nan Lv, Yulin Fan, Yongchao Hou and Xinsheng Shi were employed by the company China National Heavy Machinery Research Institute Co., Ltd. The remaining authors declare that the research was conducted in the absence of any commercial or financial relationships that could be construed as a potential conflict of interest.

References

1. Porter, T.; Findley, K.; McMurtrey, M. Assessment of creep-fatigue behavior of alloy 709. *Trans. Am. Nucl. Soc.* **2017**, *117*, 559–562.
2. Lall, A.; Bowen, P.; Rabiei, A. A study on the creep behavior of alloy 709 using in-situ scanning electron microscopy. *Mater. Charact.* **2022**, *183*, 111587. [CrossRef]
3. Zhao, Y.; Schoell, R.; Zheng, C.; Cinbiz, M.N.; Frost, M.; An, K.; Kaoumi, D. Creep properties of advanced austenitic steel 709 determined through short experiments under in-situ neutron diffraction followed by TEM characterization. *Mater. Charact.* **2021**, *182*, 111519. [CrossRef]
4. Alsmadi, Z.Y.; Alomari, A.; Kumar, N.; Murty, K. Effect of hold time on high temperature creep-fatigue behavior of Fe–25Ni–20Cr (wt.%) austenitic stainless steel (Alloy 709). *Mater. Sci. Eng. A* **2020**, *771*, 138591. [CrossRef]
5. Sourmail, T. Precipitation in creep resistant austenitic stainless steels. *Mater. Sci. Technol.* **2001**, *17*, 1–14. [CrossRef]
6. Kruizenga, A.M.; Gill, D.D.; LaFord, M.E. *Materials Corrosion of High Temperature Alloys Immersed in 600C Binary Nitrate Salt*; Sandia National Lab. (SNL-CA): Livermore, CA, USA, 2013.
7. Lv, X.; Chen, S.; Wang, Q.; Jiang, H.; Rong, L. Temperature dependence of fracture behavior and mechanical properties of AISI 316 austenitic stainless steel. *Metals* **2022**, *12*, 1421. [CrossRef]
8. Zhou, P.; Yang, W.; Wu, Y.; Zong, Y. Characterization of microstructural evolution with pre-strain in BG801 bearing steel: Grain, carbides, retained austenite and martensite. *Vacuum* **2023**, *216*, 112354. [CrossRef]
9. Ren, W.; Wang, L. Precipitation behavior of M23C6 in high nitrogen austenitic heat-resistant steel. *J. Alloys Compd.* **2022**, *905*, 164013. [CrossRef]
10. Jiao, S.; Zhang, M.; Zheng, L.; Dong, J. Investigation of carbide precipitation process and chromium depletion during thermal treatment of Alloy 690. *Metall. Mater. Trans. A* **2010**, *41*, 26–42. [CrossRef]
11. Sahlaoui, H.; Makhlouf, K.; Sidhom, H.; Philibert, J. Effects of ageing conditions on the precipitates evolution, chromium depletion and intergranular corrosion susceptibility of AISI 316L: Experimental and modeling results. *Mater. Sci. Eng. A* **2004**, *372*, 98–108. [CrossRef]
12. Pint, B.A.; Peraldi, R.; Maziasz, P.J. The use of model alloys to develop corrosion-resistant stainless steels. In Proceedings of the Materials Science Forum, Les Embiez, France, 16–21 May 2004; pp. 815–822.
13. Opila, E.J. Volatility of common protective oxides in high-temperature water vapor: Current understanding and unanswered questions. In Proceedings of the Materials Science Forum, Les Embiez, France, 16–21 May 2004; pp. 765–774.
14. Viswanathan, R.; Bakker, W. Materials for ultrasupercritical coal power plants—Turbine materials: Part II. *J. Mater. Eng. Perform.* **2001**, *10*, 96–101. [CrossRef]
15. Viswanathan, R.; Coleman, K.; Rao, U. Materials for ultra-supercritical coal-fired power plant boilers. *Int. J. Press. Vessel. Pip.* **2006**, *83*, 778–783. [CrossRef]
16. Yang, T.-X.; Li, Z.-X.; Zhou, C.-J.; Xu, Y.-C.; Dou, P. Effects of Zr and/or Ti addition on the morphology, crystal and metal/oxide interface structures of nanoparticles in FeCrAl-ODS steels. *J. Nucl. Mater.* **2023**, *585*, 154613. [CrossRef]
17. Yan, X.; Li, Z.; Lei, P.; Wang, S.; Gao, R. Significant suppression of helium bubbles in oxide dispersion strengthened FeCrAl alloys irradiated by high concentration of helium. *Vacuum* **2024**, *221*, 112915. [CrossRef]
18. Brady, M.P.; Magee, J.; Yamamoto, Y.; Helmick, D.; Wang, L. Co-optimization of wrought alumina-forming austenitic stainless steel composition ranges for high-temperature creep and oxidation/corrosion resistance. *Mater. Sci. Eng. A* **2014**, *590*, 101–115. [CrossRef]
19. Brady, M.; Wright, I.; Gleeson, B. Alloy design strategies for promoting protective oxide-scale formation. *Jom* **2000**, *52*, 16–21. [CrossRef]
20. Brady, M.P.; Yamamoto, Y.; Santella, M.L.; Pint, B.A. Effects of minor alloy additions and oxidation temperature on protective alumina scale formation in creep-resistant austenitic stainless steels. *Scr. Mater.* **2007**, *57*, 1117–1120. [CrossRef]
21. Xu, X.; Zhang, X.; Chen, G.; Lu, Z. Improvement of high-temperature oxidation resistance and strength in alumina-forming austenitic stainless steels. *Mater. Lett.* **2011**, *65*, 3285–3288. [CrossRef]
22. Sharifikolouei, E.; Sarac, B.; Micoulet, A.; Mager, R.; Watari-Alvarez, M.; Hadjixenophontos, E.; Burghard, Z.; Schmitz, G.; Spatz, J.P. Improvement of hardness in Ti-stabilized austenitic stainless steel. *Mater. Des.* **2022**, *223*, 111242. [CrossRef]
23. Farahat, A.I.Z.; El-Bitar, T. Effect of Nb, Ti and cold deformation on microstructure and mechanical properties of austenitic stainless steels. *Mater. Sci. Eng. A* **2010**, *527*, 3662–3669. [CrossRef]

24. Li, L.; Ding, Q.; Zhang, Z.; Li, Y.; Wei, X.; Zhang, Z.; Bei, H. The effect of Al and Ti on the microstructure, mechanical properties and oxidation resistance of γ'-Ni3 (Al, Ti) strengthened austenitic stainless steels. *J. Mater. Res. Technol.* **2023**, *24*, 4650–4660. [CrossRef]
25. Inoue, Y.; Hiraide, N.; Hayashi, A.; Ushioda, K. Effect of titanium on oxidation behavior of high-purity ferritic stainless steel. *Mater. Trans.* **2019**, *60*, 1968–1976. [CrossRef]
26. Pardo, A.; Merino, M.; Coy, A.; Viejo, F.; Carboneras, M.; Arrabal, R. Influence of Ti, C and N concentration on the intergranular corrosion behaviour of AISI 316Ti and 321 stainless steels. *Acta Mater.* **2007**, *55*, 2239–2251. [CrossRef]
27. Qi, W.; Guoshuai, C.; Zhangjian, Z.; Ru, X.; Jiyun, Z.; Rui, T.; Lefu, Z. Effect of N and Al Addition on Microstructure and Mechanical Properties of Modified 25Ni-20Cr Austenitic Stainless Steel Aged at 700 °C. *Nucl. Power Eng.* **2023**, *44*, 275–283.
28. Morales, M.; Rezayat, M.; García-González, S.; Mateo, A.; Jiménez-Piqué, E. Ru-$Ce_{0.7}Zr_{0.3}O_{2-\delta}$ as an anode catalyst for the internal reforming of dimethyl ether in solid oxide fuel cells. *Nanomaterials* **2024**, *14*, 603. [CrossRef]
29. Taneike, M.; Abe, F.; Sawada, K. Creep-strengthening of steel at high temperatures using nano-sized carbonitride dispersions. *Nature* **2003**, *424*, 294–296. [CrossRef] [PubMed]
30. Hu, J.; Xiao, Z.; Huang, Y. Segregation of solute elements and their effects on the strength of Al Σ5 (210) [001] symmetrical tilt grain boundary in 2219 alloys. *Mater. Sci. Eng. A* **2021**, *800*, 140261. [CrossRef]
31. Peng, H.; Huo, W.; Zhang, W.; Tang, Y.; Zhang, S.; Huang, L.; Hou, H.; Ding, Z.; Liu, F. Correlation between stabilizing and strengthening effects due to grain boundary segregation in iron-based alloys: Theoretical models and first-principles calculations. *Acta Mater.* **2023**, *251*, 118899. [CrossRef]
32. Li, D.; Feng, Y.; Song, S.; Liu, Q.; Bai, Q.; Wu, G.; Lv, N.; Ren, F. Influences of Nb-microalloying on microstructure and mechanical properties of Fe–25Mn–3Si–3Al TWIP steel. *Mater. Des.* **2015**, *84*, 238–244. [CrossRef]
33. Wang, H.; Du, H.; Wei, Y.; Hou, L.; Liu, X.; Wei, H.; Liu, B.; Jia, J. Precipitation and properties at elevated temperature in austenitic heat-resistant steels—A review. *Steel Res. Int.* **2021**, *92*, 2000378. [CrossRef]
34. Qu, S.; Huang, C.; Gao, Y.; Yang, G.; Wu, S.; Zang, Q.; Zhang, Z. Tensile and compressive properties of AISI 304L stainless steel subjected to equal channel angular pressing. *Mater. Sci. Eng. A* **2008**, *475*, 207–216. [CrossRef]
35. Uchic, M.D.; Dimiduk, D.M.; Florando, J.N.; Nix, W.D. Sample dimensions influence strength and crystal plasticity. *Science* **2004**, *305*, 986–989. [CrossRef]
36. *A240/A240M-19*; Standard Specification for Chromium and Chromium-Nickel Stainless Steel Plate, Sheet, and Strip for Pressure Vessels and for General Applications. ASTM International: West Conshohocken, PA, USA, 2019.
37. Calmunger, M.; Chai, G.; Eriksson, R.; Johansson, S.; Moverare, J.J. Characterization of austenitic stainless steels deformed at elevated temperature. *Metall. Mater. Trans. A* **2017**, *48*, 4525–4538. [CrossRef]
38. Zhao, Y.; Cinbiz, M.N.; Park, J.-S.; Almer, J.; Kaoumi, D. Tensile behavior and microstructural evolution of a Fe-25Ni-20Cr austenitic stainless steel (alloy 709) from room to elevated temperatures through in-situ synchrotron X-ray diffraction characterization and transmission electron microscopy. *J. Nucl. Mater.* **2020**, *540*, 152367. [CrossRef]
39. Saenarjhan, N.; Kang, J.-H.; Kim, S.-J. Effects of carbon and nitrogen on austenite stability and tensile deformation behavior of 15Cr-15Mn-4Ni based austenitic stainless steels. *Mater. Sci. Eng. A* **2019**, *742*, 608–616. [CrossRef]
40. Shim, J.-H.; Kozeschnik, E.; Jung, W.-S.; Lee, S.-C.; Kim, D.-I.; Suh, J.-Y.; Lee, Y.-S.; Cho, Y.W. Numerical simulation of long-term precipitate evolution in austenitic heat-resistant steels. *Calphad* **2010**, *34*, 105–112. [CrossRef]
41. Kumar, B.R.; Mahato, B.; Bandyopadhyay, N.; Bhattacharya, D.K. Comparison of rolling texture in low and medium stacking fault energy austenitic stainless steels. *Mater. Sci. Eng. A* **2005**, *394*, 296–301. [CrossRef]
42. Das, A.; Tarafder, S. Experimental investigation on martensitic transformation and fracture morphologies of austenitic stainless steel. *Int. J. Plast.* **2009**, *25*, 2222–2247. [CrossRef]

nanomaterials

Review

A Comprehensive Review of High Burn-Up Structure Formation in UO_2: Mechanisms, Interactions, and Future Directions

Zhenhong Ge, Dong Yan, Penghui Lei * and Di Yun *

School of Nuclear Science and Technology, Xi'an Jiaotong University, Xi'an 710049, China; gezhenhong@stu.xjtu.edu.cn (Z.G.); flavia990331@stu.xjtu.edu.cn (D.Y.)
* Correspondence: penghuilei@xjtu.edu.cn (P.L.); diyun1979@xjtu.edu.cn (D.Y.)

Abstract: In the rim zone of UO_2 nuclear fuel pellets, high burn-up and low temperatures drive changes in the microstructure, leading to the formation of high burn-up structures (HBS). This review focuses on the formation of HBS, beginning with a description of the two contentious mechanisms—recrystallization and polygonization—that are believed to be the primary controlling factors. We discuss experimental and simulation studies that support both mechanisms, emphasizing that although each mechanism can explain certain aspects of HBS formation, neither recrystallization nor polygonization alone is sufficient to fully explain the observed phenomena. Furthermore, we emphasize the intrinsic relationship between these two mechanisms, suggesting that they represent different manifestations of the same underlying process under varying conditions, and we reference relevant studies that support this perspective. Lastly, we underline the significance of investigating the formation processes of HBS and provide an outlook on future research directions based on the current state of knowledge.

Keywords: uranium dioxide; high burn-up structure; recrystallization; polygonization

1. Introduction

Due to the neutron energy spectrum configuration in pressurized water reactors (PWRs), the intense radiation self-shielding effect can cause the UO_2 fuel pellet near the edge to experience local burn-up exceeding the average burn-up by more than double. When the local burn-up reaches around 50 GWd/tHM, a noticeable grain refinement process begins, resulting in a new fuel morphology. By the time the fuel burn-up reaches approximately 75 GWd/tHM, this transformation is nearly complete. This newly formed structure is referred to as HBS, also known as the rim structure, rim effect, or described as "cauliflower-like" [1] based on its morphology.

It is generally accepted that HBS exhibit the following characteristics:

(1) The original grains, typically measuring around 10 μm in the fresh fuels, are subdivided into 10^4–10^5 new grains, each measuring 100–300 nm;
(2) The appearance of micron-sized intergranular fission gas pores, indicating a larger porosity in the HBS;
(3) The depletion of Xe and dislocations in the newly formed fine-grained matrix.

The HBS occupy only a thin region at the radial edge of the fuel and expand inward radially as burn-up increases. Some researchers [2] believe that under suitable conditions, this expansion can extend deeply into the fuel. Moreover, the formation of such HBS has been observed not only in UO_2 fuel but also in Pu-rich mixed U-Pu fuels (MOX) and U-Mo fuels [3]. Similar grain refinement phenomena, potentially indicative of HBS formation, have also been observed in carbide [4] and silicide fuels [5].

As early as the 1950s, localized restructuring phenomena were noted in naval reactors [6] and were observed in commercial reactors in the early 1960s [7]. However, with the ongoing development of commercial reactors aimed at alleviating the rapid increase of spent fuel and enhancing economic efficiency through higher operational burn-up, the newly formed HBS have raised concerns about the safety implications of increased burn-up, especially after studies indicating degradation in thermal conductivity [8] and mechanical properties [9] associated with HBS.

Thus, the importance of measuring and modeling HBS to assess their impact during operation has become increasingly prominent. This raises the question of how HBS are formed, and the implications of this formation process are unavoidable for common burn-up levels in commercial PWRs. In the following sections, we will discuss research on the formation of HBS and present some insights.

2. The Formation Mechanism of HBS: An Unresolved Issue

Due to the complexities of nuclear fuel systems and the numerous influencing factors, coupled with the harsh service environments and objective limitations of post-processing, the application of characterization techniques is significantly restricted. These combined factors result in a lack of systematic and comprehensive scientific descriptions of the formation of HBS. Existing experimental results can only focus on microstructural morphology and elemental distribution in localized areas at specific burn-ups, without capturing the continuous changes in microstructural characteristics at the same location under varying burn-up conditions. The prolonged cooling time after removal from the reactor introduces errors, which notably affect the observation of the microstructural features. Furthermore, experimental constraints in early characterization further limit the observation of this process.

There is a widespread debate regarding the mechanisms behind HBS formation, with two prominent perspectives: one posits that HBS arise from the formation and growth of subgrains into recrystallization nuclei, followed by the growth of recrystallized grains, representing a recrystallization mechanism, as shown in Figure 1. The other suggests that they result from the entanglement of dislocations forming dislocation walls, which then rearrange, leading to the subdivision of original grains, representing a polygonization mechanism, as shown in Figure 2. Both mechanisms are supported by relevant experimental evidence, such as the presence of numerous nano-grains with high-angle boundaries, generally regarded as features of recrystallization, observed in edge-region EPMA characterizations, as well as the newly formed relatively large grains exhibiting low-angle boundaries, which could only arise from polygonization.

Discussions based on these mechanisms continue to emerge, along with preliminary explorations of evolutionary process models and their impacts based on experimental characterizations. The establishment of these models also reflects the ongoing debate surrounding these mechanisms. Since the review by Rondinella et al. [5] has already provided a thorough summary of earlier literature, this overview will primarily focus on research related to HBS conducted after the year 2000.

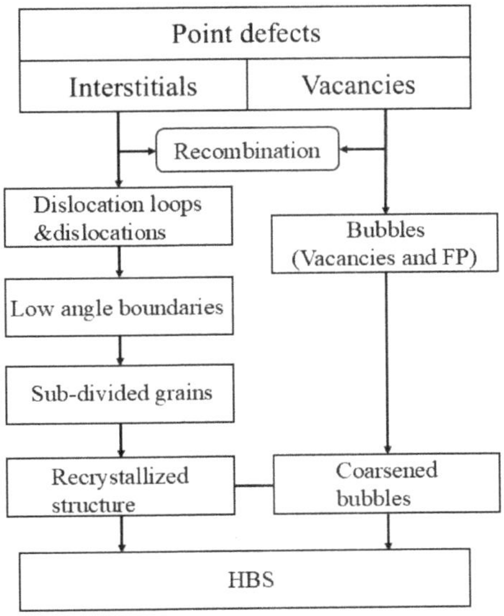

Figure 1. The recrystallization mechanism proposed by Nogita and Une et al. [10].

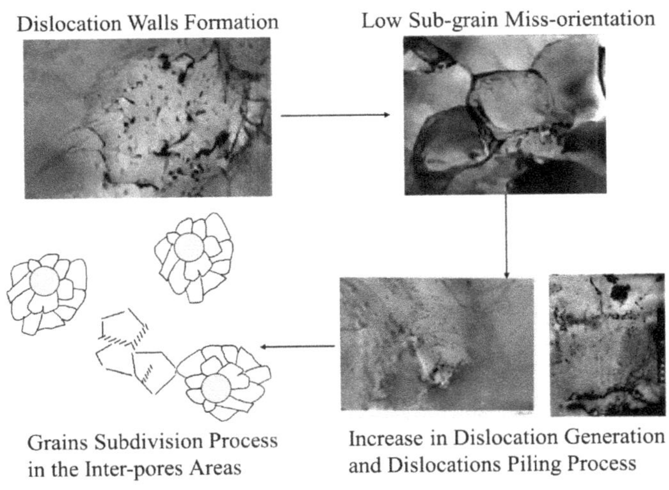

Figure 2. The polygonization mechanism proposed by Baron et al. [2].

2.1. *Observation from the Experimental Perspective*

Nogita and Une [10,11] proposed that at lower burn-up (44 GWd/tU), the tangled dislocation network with low-angle grain boundaries is formed by the inhomogeneous accumulation of dislocations. In contrast, for higher burn-up fuels, the energy stored in the matrix is released, resulting in the original grains being subdivided into high-angle grain boundary grains of 20–30 nm size, which serve as the nuclei for subsequent irradiation-induced recrystallization. The strain energy in the matrix, driven by the accumulation of irradiation defects, facilitates recrystallization, and past research work indicates that temperature has a significant impact on this process. They also noted that the recrystallization process eliminates intragranular bubbles and dislocations, asserting that recrystallized grains do not undergo secondary recrystallization. Furthermore, the relationship between dislocation density and burn-up was summarized (although the temperature effect is relatively limited, especially within the lower temperature range at the periphery): $\log(\rho_d) = 2.2 \times 10^{-2}\beta + 13.8$. This relationship is widely utilized in phase-field simulations to change the free energy. Une further noted in experiments [12] with large grain-sized UO_2 fuel that larger grain sizes significantly enhance resistance to HBS. It was also observed that high-density dislocations and subdivided grains first appear at the grain boundary locations, providing additional support for the proposed HBS formation mechanism.

Spino et al. [13] used a specially developed micro X-ray diffraction technique to measure the radial variations in lattice parameters and peak widths in two high burn-up UO_2 fuels (67 and 80 GWd/tM). They found that α-damage decreases both toward the hotter center and the cooler edge of the pellet. The reduction toward the center was attributed to thermal annealing, while the reduction toward the pellet edge was linked to the formation of the HBS. They also noted that the formation of the HBS requires significant lattice damage and the accumulation of xenon gas atoms trapped within the matrix.

Noirot et al. [14] concluded from experimental observations that the formation of HBS is driven by fission products, rather than fission itself. They suggested that there is a threshold temperature, above which subgrains do not form. Their observations of samples at 45 GWd/tU revealed that HBS appear within grains, and the impact of grain size on grain boundaries may have a limited role in inhibiting HBS formation. Based on the experimental evidence, they proposed that HBS initiation is primarily triggered by the solid solution of trivalent rare earths and plutonium, rather than by gaseous, metallic, or volatile fission products. However, they also noted that subsequent changes involving bubble formation indicate that these other fission products contribute to the further development of HBS. Further examination [15] of the M5-clad fuel irradiated to 83 GWd/tU indicated that the fuel porosity gradually increased between fuel radial locations of 0.43R and 1R, where R denotes the fuel radius, with only a limited increase in the range of 0.43R to 0.57R. It is noteworthy that the large, irregular (non-spherical and far from lens-shaped) bubbles formed at the center of the fuel retained most of the unreleased fission gases, with the majority of these bubbles located at the boundary [16].

Baron et al. [2] proposed that the development of the HBS occurs in three sequential stages: the construction of a periodic dislocation network, the loss in EPMA gas detection, and the coalescence and growth of intragranular nanoscale bubbles followed by grain refinement. Based on experimental observations, with burn-ups above 55 GWd/tU, they suggested that this grain refinement is predominantly driven by the polygonization mechanism, which contradicts the perspective of Nogita and Une et al. [10,11]. Additionally, they speculated that the propagation of grain subdivision may be caused by the punching effect due to pressurized pores or the high stress in the regions between pores.

T. Sonoda et al. [17] characterized 16 fuel samples irradiated at temperatures ranging from 400 °C to 1300 °C, with burn-ups of 36 to 96 GWd/tU. Building upon the aforementioned experiments, they further expanded the temperature and burn-up ranges in their study and suggested that both temperature and burn-up play a critical role in the formation of HBS. They observed that the HBS transformation threshold occurred between 55 and 82 GWd/tU, with a temperature threshold of 1100 ± 100 °C. TEM analysis revealed that most subgrain boundaries were of low-angle type, and 0.5–2 μm polygonal grains were also present, strongly supporting the polygonization mechanism. In high-temperature irradiated samples, dislocations appeared to be pinned by fission products, indicating a possible pinning effect. They further conducted ion irradiation experiments [18] on UO_2 using 100 MeV Zr^{10+} and 210 MeV Xe^{14+} ions to study the properties of ion tracks and the microstructural evolution under ion track accumulation. The experimental results indicated that ion irradiation leads to shape changes in fabricated pores (from spherical to elliptical) and dislocation generation. The significant changes in both the surface morphology and internal structure of UO_2 suggested that the overlapping of ion tracks enhances the diffusion of point defects and dislocations, facilitating subgrain formation at relatively low temperatures. The study also suggested that subgrain formation could occur independently of fission product presence.

In the New Cross-Over project (NXO), Kinoshita et al. [19,20] used Xe ion implantation in CeO_2, which has a similar fluorite structure to UO_2, to recreate subgrain structures analogous to those found in high burn-up fuels. Through TEM and synchrotron radiation XRD, they determined that oxygen plays a crucial role in the reorganization and recovery of nanoscale crystals. Additionally, computational results from first-principles calculations [19,21], molecular dynamics [20], and accelerated molecular dynamics [19,22,23] further supported the idea that oxygen atom clusters and the resulting stable structures may play a significant role in the formation of HBS.

Miao et al. [24] irradiated microcrystalline UO_2, produced by plasma sintering, with 84 MeV ions at 350 °C to a dose of 1357 dpa, discovering the presence of nano-grains and pores characteristic of HBS. They conducted a quantitative study on the sizes and number densities of the formed bubbles as well as the grain size distribution, analyzing the misorientation of the grain boundaries. Their findings provide partial support for the perspectives proposed by Baron et al. [2] regarding the formation process of HBS. It was found that low-angle grain boundaries predominated (83%), leading to the conclusion that the formation mechanism of HBS is a polygonization process driven by radiation-induced dislocations associated with pipe and grain boundary diffusion of Xe and vacancies.

Baranov et al. [25] conducted ion irradiation of SIMFUEL fuel in the energy range of 20 keV (He^+) to 320 keV (Xe^{16+}) and 1–90 MeV (Xe^{24+}). They observed grain refinement only under high-energy ion irradiation. They posited that, for low-energy ions, the size of the coherent scattering regions decreases with increasing dislocation density, which limits dislocation motion and subsequently restricts the formation of subgrains. In contrast, high-energy ions, upon reaching critical conditions, increase the size of the coherent scattering regions, allowing for dislocation rearrangement and enhancing dislocation diffusion creep, thereby promoting grain refinement based on the polygonization mechanism. They also noted that dispersed particles, represented by small bubbles, can hinder dislocation movement, further limiting grain refinement.

Gerczak et al. [26] analyzed the nearly complete cross-section of high burn-up commercial fuel with local burn-ups of 60 to 184 GWd/tU. The fuel experiences significant microstructural variations from the fuel-cladding interface to the fuel pellet center, influenced jointly by temperature and local burn-up, which aligns with the structural changes initially observed by Spino et al. [13]. By examining grain boundary misorientation and size, they identified a transition region near fully developed HBS and suggested that the grain refinement is primarily driven by polygonization mechanisms due to the formation of low-angle grain boundaries. They also speculated that the disappearance of low-angle boundary features within the region transitioning to HBS is caused by creep deformation in the pellet. Cappia et al. [27] irradiated UO_2 samples with original grain sizes of 7.2/10 μm at 623/723 K using 84 MeV Xe^{26+} ions to doses of 1357/1493 dpa. They also observed that low-angle grain boundaries dominated the subgrain boundaries. However, in the regions where grain refinement occurred, no bubbles were found, leading them to conclude that the presence of large bubbles is not a necessary condition for grain refinement. This contradicts the viewpoint of Baron et al. [2].

Hiernaut's annealing experiments [28] on high burn-up samples with a local burn-up of 220 ± 20 MWd/kgHM showed that a significant portion of the fission gases remains trapped within the HBS. The pressure within the HBS pores can reach up to 30 MPa, causing grains to fracture along the original grain boundaries. This indicates that overpressure may be a contributing factor in the formation of the HBS.

Xiao et al. [29] irradiated samples under conditions of linear power ranging from 260 to 300 W/cm and temperatures between 300 and 400 °C, achieving burn-up depths of 48–120 GWd/tU. High-resolution scanning electron microscopy (SEM) was used to observe fuel samples with varying local burn-up (at different radial positions) within the 48–120 GWd/tU range. Observations of low burn-up samples revealed the formation of subgrains with low-angle boundaries away from coarsening bubbles, which is consistent with the findings of Baron et al. [2] on grain refinement. As burn-up increased, small grains with high-angle grain boundaries appeared around the bubbles, alongside the formation of new low-angle subgrains. In the highest burn-up samples, a greater number of high-angle grain boundaries surrounded the coarsening bubbles. Based on these findings, they concluded that both polygonization and recrystallization are mechanisms contributing to the formation of HBS.

Barker et al. [30] examined approximately 30 AGR samples exhibiting HBS after irradiation using SEM and light optical microscopy (LOM), along with similarly sized AGR fuel samples without HBS. They observed that the minimum burn-up for the emergence of HBS was 27 GWd/tU, suggesting that this earlier threshold burn-up may be influenced by the moderator. They also noted that the early stages of HBS formation seem to initiate within the grains, indirectly supporting the polygonization mechanism.

Cappia et al. [31] conducted a study on the microstructural features and volatile fission products of high burn-up samples with local burn-up reaching 76 GWd/tHM across different radial positions. Experimental observations revealed that irradiation introduced a significant number of defect damages (this observation is consistent with the findings of T. Wiss et al. [32]), resulting in the formation of a dislocation network. From a microstructural perspective, three distinct regions were identified along the radial position:

1. In the central region, noticeable polygonization occurred, resulting in subgrain domains separated by low-angle grain boundaries;
2. In the region corresponding to a relative radius of 0.55 to 0.8, there were no clear features of grain subdivision, lower porosity compared to the central area, and a high retention of nano-sized bubbles in the matrix;
3. At the outer region, near a relative radius of 0.98, clear HBS characteristics were observed.

Zacharie-Aubrun I et al. [33] characterized standard and Cr-doped UO_2 samples within a burn-up range of 35–73 GWd/tU using EBSD techniques. They observed distinct HBS at the fuel edges, which was independent of the initial microstructure. For all types of samples, when the burn-up exceeded 61 GWd/tU, a restructuring phenomenon was evident in the central region, believed to be related to the formation of bubbles. Furthermore, they noted that this restructuring intensified with increasing burn-up, and the orientation of the grains formed in the restructuring area was similar to that of the original grains.

Based on the aforementioned experimental studies, the formation conditions of HBS in UO_2, including temperature, burn-up threshold, and the original grain size—an important influencing factor identified in numerous experiments—along with the proposed formation mechanisms and fuel types, are summarized in Table 1. Here, a single temperature value corresponds to the burn-up threshold for the formation of the HBS at that temperature.

Table 1. Formation conditions of HBS from the experimental perspective (blank indicates values not provided).

Temperature (°C)	Burn-Up (GWd/tU)	Original Grain Size (μm)	Reactor, Type	Supporting Views	Researcher (Ref.)
400	60–70		BWR, rod	recrystallization	Nogita/Une [10,11]
<800	70	17	PWR, rod	both	Spino [13]
	45 **	10	PWR, rod	polygonization	Noirot [14]
<1100	55–65			polygonization	Baron [2]
<1100 ± 100	55–82	9.7 ± 0.8	disk fuel	polygonization	T. Sonoda [17]
550 *	100 *	~8	disc samples	polygonization	Gerczak [26]
	220 ± 20 *		PWR, rod		Hiernaut [28]
300–400	87	~8	RR, disk fuel	both	Xiao [29]
~750	27		AGCR, rod		Baker [30]
<1200 *	76 *		LWR, rod		Cappia [31]
<1000 *	61	~9	PWR, rod	polygonization	Zacharie-Aubrun I [33]

Note: * according to the end-state data provided in the paper. ** average burn-up.

2.2. Insights Derived from Numerical Simulations

In the simulation study of HBS formation, an attempt was first made to derive the required kinetic parameters for the evolution of the kinetic equations by combining experimental data and theoretical assumptions. The evolution of point defects, dislocations, bubbles, and other defects with burn-up at different temperatures was investigated. For instance, in the treatment of fission gas bubbles, the following approach is commonly adopted [34]:

$$\frac{dC_{\alpha,i}}{dt} = -a_{\alpha,i}C_{\alpha,i}^2 - b_{\alpha,i}C_{\alpha,i} + e_{\alpha,i} \quad (1)$$

where i represents the size class of the bubble, which is distinguished according to the number of gas atoms contained in a bubble; α represents the location of the bubble: lattice, dislocation, grain face, and grain edge. $C_{\alpha,i}$ represents the number density of the i-th size class of bubbles at the α position (when i = 1, the number density corresponding to the fission gas atoms is changed to C_g); $a_{\alpha,i}$, $b_{\alpha,i}$, $e_{\alpha,i}$, are the rate coefficients representing different fission gas behaviors affecting the bubble number density, respectively. Based on these computational results mentioned before, a relationship between the computed values and the experimentally observed threshold for HBS formation was established. This criterion allows for effective simulation of the restructuring process.

Martín Lemes et al. [35] observed a significant change in behavior around 100 MWd/kgU based on experimental data [36,37] regarding porosity and pore number density changed with burn-up: the rate of increase in porosity slowed down, while the pore number density reached a maximum before declining. Consequently, high burn-up was categorized into two cases: 60–100 MWd/kgU and greater than 100 MWd/kgU (defined by Romano A [37] as ultra HBS. Using empirical hydrostatic pressure equations that vary with burn-up and a modified Van der Waals equation, along with established empirical equations for porosity and pore number density, they studied the fission gas overpressure and the retention of fission gases in the matrix and pores under the two cases separately. The results compared well with existing published data. This threshold is confirmed by the pore size study of fuels with a burn-up of 80–200 MWd/kgU by Cappia et al. [38].

Rest and Hofman analyzed grain refinement based on the principles of metal recrystallization [39,40]. Their theory posits that during irradiation, cellular dislocation structures form at an early stage. Impurities generated during fission subsequently diffuse to the void dislocation structures as vacancy-impurity complexes, thereby hindering dislocation motion across the walls. The impurity-free outer walls further undergo subgrain coalescence. The core hypothesis for the recrystallization process is to control the concentration of unit volume nuclei through kinetic equations. As the fission process progresses, energy becomes increasingly concentrated in fewer nuclei. Grain refinement occurs when the creation of grain boundary surfaces is offset by the creation of strain-free volume, resulting in a net decrease in the material's free energy.

The model subsequently considers the effects of impurity pinning and fission gas bubbles on the recrystallization nuclei [41]. Furthermore, based on observations and modeling of the U-Si compound system [42,43], the authors incorporate the amorphous clusters generated by collision cascades and the centers of compression produced by "shock waves" into the model as influencing factors for nuclei control. Finally, using the same theory, they extend this recrystallization simulation to U-xMo fuel with similar phenomena [44], explaining the depletion of Xe, pore formation [45], and swelling [46] effects influenced by the recrystallization process. Rest also noted that the model is supported by the experimental observations of submicron recrystallized grains in 100 GWd/tU fuel made by Thomas et al. [47] and Nogita et al. [10].

Veshchunov et al. [48] developed and implemented a model for dislocation generation and evolution under irradiation conditions in the MFPR framework, combining it with a set of equations for point defect evolution and their interactions with bubbles. This model was validated against the experimental data of Nogita [11]. Formation and growth of intragranular fission gas bubbles in UO_2 fuels with burn-up of 6–83 GWd/tU provides a self-consistent consideration of the entire system in irradiated fuel, including point defects (vacancies, interstitial atoms, and gas atoms) as well as extended defects (bubbles, dislocations, vacancy loops, and pores). The model predicts a significant reduction in dislocation density at temperatures between 1300 K and 1400 K. Furthermore, at burn-ups of at least 100 GWd/tU, dislocation density does not reach the threshold, preventing grain restructuring. This is consistent with the observations of Kinoshita et al. [20]. Based on this, it is further suggested that the dislocation density may be a key factor influencing the temperature threshold for restructuring of HBS.

Xiao et al. [49] developed a model referencing the mechanisms proposed by Rest [39,40], which incorporates dislocation evolution under irradiation and the successive grain subdivision process as local burn-up increases. They hypothesize that the primary driving force for grain subdivision in the edge regions of high burn-up UO_2 fuel matrices is the generation and accumulation of dislocation loops under irradiation. Based on this, they predicted the sizes of subdivided grains at different burn-up levels (50–135 GWd/tU) and dislocation density (6×10^{16}–2.7×10^{17} m^{-2}), and the simulation results showed good agreement with the experimental data [25]. The calculation results for dislocation density in the temperature range of 900–1400 K indicate that the dislocation density of as-irradiated UO_2 is highly insensitive to temperature changes and decreases fleetly within this range. It is further noted that grain refinement is likely to occur in the temperature range of 1300–1400 K only when the burn-up reaches at least 100 GWd/tU, which is consistent with experimental observations [17].

D. Pizzocri et al. [50] did not discuss the mechanisms of polygonization or recrystallization. Instead, they based their work on the concept of effective burn-up (which considers radiation damage accumulation only when the temperature is low enough to avoid thermal recovery of radiation damage) and measured grain sizes from experimental data. They directly fitted a first-order differential model to obtain the evolution equation for grain size as a function of burn-up, which aligns well with the experiments conducted by Ray [51] and Spino [36].

The phase field method, driven by the reduction of free energy, can simulate the microstructural evolution process at the mesoscopic scale, allowing for the study of microstructural morphology and the effects of external fields (such as thermal, mechanical, and irradiation). The evolution of HBS is described using a governing equation similar to the following form [52], which accounts for changes in energy:

$$F = \int_V [f_{chem} + f_{poly} + f_{st} + W\, g(\eta) + f_{grad}(\eta, \varphi)] dV \tag{2}$$

where f_{bulk} represents the chemical free energy density, f_{poly} is the polycrystalline energy density, f_{st} is the stored strain energy density, and f_{grad} is the gradient energy density, respectively. $g(\eta)$ is the double well function and W is the potential height.

A modified Cahn–Hilliard equation is employed to describe the concentration changes of conservative field variables, such as vacancy and gas atom concentrations. For example, the governing equation for fission gas atom evolution is expressed as

$$\frac{\partial c_g}{\partial t} = \nabla \cdot M_v \nabla \mu_v + P_g \tag{3}$$

where M_g is the mobilities of Xe gas atom with the form and P_g is the source term introducing gas atoms in the matrix phase.

The nucleation probability is calculated in a manner related to dislocation density (burn-up), thereby introducing nano-grains, which subsequently influence the free energy changes and simulate the evolution of the microstructure. Over the past decade, it has been widely applied in the field of microstructural evolution of nuclear fuel.

Abdoelatef et al. [53,54] established a phase field model on the MOOSE platform, considering interfacial energy at grain boundaries and bubble surfaces, dislocation-related strain energy, and the chemical energy of gas atoms to describe the formation and evolution of HBS. The simulation results indicate that dislocation density, grain boundary energy, bubble radius, and number density significantly influence the recrystallization fraction, grain count, and average grain size.

K. Ahmed et al. [55] addressed the recrystallization nucleation problem by considering simplified Langevin-type fluctuations in the evolution equations. They derived a critical dislocation density threshold for recrystallization: when the dislocation density exceeds this threshold, recrystallized grains grow; otherwise, they shrink. The evolution of dislocation density is also based on an empirical relationship [10] that depends solely on burn-up. The model further developed [56] has been shown to successfully simulate both homogeneous and heterogeneous nucleation of bubbles and subgrains. The computed subgrain size results align well with experimental findings [49] at 1200 K.

Amani Cheniour et al. [57] utilized the MARMOT based on the MOOSE platform and classical nucleation theory to simulate recrystallization at fixed core sizes under typical conditions of commercial light water reactor fuel and accelerated conditions at 650/800 °C and 40–60/40–90 MWd/kgU. The simulation results revealed significant differences in the variation of recrystallization fraction and starting point position, highlighting the notable impacts of power and temperature.

Permann et al. [58] introduced recrystallization grains using a random nucleation approach and established the concept of effective dislocation density based on the mechanism by which recrystallization eliminates dislocations. They used the formula proposed by Staker [59] to establish a relationship between the newly introduced grains and dislocation density, simulating the size and quantity of recrystallized grains based on the MOOSE platform.

Y. Jiang et al. [60] studied the impact of recrystallization grains on bubbles in UMo through random nucleation. The increase in grain boundary area fraction from recrystallization accelerated the formation and growth of bubbles. They further applied the explicit dynamic nucleation theory proposed by R. Ding [61] to introduce a nucleation rate for HBS of UO_2 [52], calculating the nucleation rate per unit grain boundary area in the Arrhenius form. This successfully described the three-stage variation in porosity under synchronous bubble evolution conditions at 673 K and burn-ups ranging from 0 to 144 GWd/tU, aligning with experimental observations [26]. They also pointed out that the recrystallization process can promote bubble evolution, with the increased grain boundary area accelerating the nucleation of recrystallized grains and bubbles.

W. Jiang et al. [62] proposed a phase field model for quasi-brittle fracture in HBS with pressurized cracks under 700 K. It is suggested that the formation of HBS induced by the recrystallization mechanism affects the grain structure, thereby altering the crack propagation path and the morphology of the resulting fragments.

Baranov [63] established a dislocation dynamics model based on four premises: (P1) dislocations generation from radiation source are slow, (P2) because fresh fuel is brittle, dislocations are fixed after their generation by the radiation source, (P3) glid mobility is unaffected by irradiation, and (P4) short-term irradiation can be considered steady-state. The study investigated the distribution of dislocations and the effects of irradiation on dislocation motion. It was noted that, under the experimentally observed dislocation density and at a temperature of 573 K, interacting dislocations exhibit a periodic structural characteristic with spacings of 100–300 nm. This supports the view that HBS are induced by polygonization mechanisms and provides predictions for the minimum subgrain size produced by polygonization.

Jonnet [64–66] assumed that the formation of HBS is a polygonization process, where dislocations are essential for the formation of new subgrain boundaries. Therefore, his research model focuses on the role of dislocation-type defects—specifically, the nucleation, growth, and accumulation of interstitial-type dislocation loops formed by the necessary shear dislocation lines and the redistribution of point defects in the fuel. Based on this approach, he analyzed the theoretical stresses induced by the presence of various possible configurations of dislocations and emphasized the growth mechanisms of interstitial-type dislocation loops. This work provides theoretical support for further studies on the mechanisms of subgrain formation.

Based on the aforementioned numerical simulations, the formation conditions of HBS in UO_2, including temperature, burn-up threshold, and the original grain size—an important influencing factor identified in numerous experiments—along with the proposed formation mechanisms, are summarized in Table 2. Here, a single temperature value corresponds to the burn-up threshold for the formation of the HBS at that temperature.

Table 2. Formation conditions of HBS from the numerical simulations (blank indicates values not provided).

Temperature (°C)	Burn-Up (GWd/tU)	Original Grain Size (μm)	Supporting Views	Researcher (Reference)
<727	~50	10~20	recrystallization	Rest [39,40]
<1027	55–82	~9	both	Veshchunov [48]
<1027–1127	100		polygonization	Xiao [49]
827	48.6		polygonization	Permann [58]
<1000	50	9–12	polygonization	D. Pizzocri [50]
927	70–75	2.9–4.1	polygonization	Abdoelatef [53,54]
927	45	6.4		K. Ahmed [55]
650	58			
800	59	10	both	Amani Cheniour [57]
827	64.2	10		Permann [58]
400	50	8.01	polygonization	Y. Jiang [60]
<1000	70			Jonnet [64–66]

In summary, both experimental and simulation studies face difficulties in determining the precise temperature and burn-up values required for the formation of HBS, with these typically being presented as ranges or estimates. Due to the limitations of experimental setups, it is challenging to precisely capture the specific threshold for HBS formation, and the observed phenomena are constrained by the sample conditions. Furthermore, the parameterization in simulation studies is heavily dependent on the experimental characterization results. The debate between these two perspectives persists, influenced by varying fuel characterization outcomes and the limitations of the implementation methods.

3. Proposal of Hybrid Control Mechanism

3.1. Theoretical Basis

The above research all put forward their own views based on the observation of the actual experimental situation and explored the establishment of the model to a certain extent. However, the aforementioned studies generally consider HBS to be local phenomena that arise under specific temperature and burn-up conditions, predominantly governed by a single mechanism. From a holistic perspective on the fuel, one can argue that both mechanisms share commonalities or, further, that they represent different manifestations of a universal mechanism under varying conditions.

Based on existing experimental observations, the following hypothesis is proposed: from an energetic perspective, the influences of both mechanisms can be considered collectively. Liang et al. [67] proposed a relationship regarding the energy of deformed grains, suggesting that the energy accumulated due to dislocations is proportional to dislocation density and the shear modulus. Due to the influences of irradiation and fission products, dislocations continuously accumulate during the irradiation process and become pinned, resulting in a significant increase in energy within the system. Concurrently, influenced by fission gases or other volatile fission products, pores continuously form during the fuel operation, with their volume affected by local power and temperature. The recovery of dislocations is significantly temperature-dependent; as temperature increases, dislocation density substantially decreases.

As mentioned in Section 2.1, Nogita et al. [10] pointed out that as burn-up increases, the dislocation density gradually accumulates. The decline in dislocation density, thereby lowering the system's energy, can only be achieved through recrystallization. In the central region of the fuel, although the temperature is higher, dislocation density is relatively low. However, from the modified Van der Waals equation, it can be inferred that higher temperatures lead to larger fission gas bubbles, which, on the one hand, adversely affect material properties and, on the other hand, pin the dislocations, further increasing the energy of the system. In the high-temperature region, the crystals rearrange dislocations to form low-angle grain boundaries, which leads to a configuration that lowers the internal energy of the system. However, when this energy has not reached the threshold necessary for the recrystallization process to occur, recrystallization cannot be initiated. This observation aligns with the previously mentioned polygonization theory.

For what is traditionally considered HBS, subgrain formation can, to some extent, be regarded as a precursor stage to the recrystallization process. Humphreys et al. [68] indicate that the nucleation stage of recrystallization occurs through the mechanism of subgrain growth, where recrystallization nucleation should take place in regions of high strain energy and high orientation gradients leading to discontinuous subgrain growth. Given that these conditions fall within the optimal temperature range for UO_2 recrystallization (where reported grain refinement commonly occurs), coupled with significant burn-up differences due to radial power discrepancies (as indicated in research by D. Pizzocri [50]), the local high dislocation density achieved is insufficient to reduce system energy below the energy threshold, prompting a series of recrystallization processes.

Due to the significant reduction in grain size occurring concurrently with recrystallization, the migration of gas atoms and bubbles within grains toward the grain boundaries is markedly enhanced. This is facilitated both by the noticeably shortened diffusion paths and the effective trapping of these processes by the pores. Additionally, this process is accompanied by the transformation of original intragranular bubbles into intergranular bubbles (potentially captured by dislocation structures), further exacerbating this migration process. This leads to the observed phenomenon of intragranular Xe gas atom depletion in the matrix, further forming pores, which enhances local energy and accelerates the recrystallization process. Further observations [33] in the fuel center region support this notion. As the burn-up increases, larger pores gradually form within the center. The potential overpressure at these pore locations, combined with dislocation accumulation, significantly raises the local energy, thereby triggering a limited recrystallization process. This seems to provide theoretical support for Baron et al.'s [69] energy-threshold-based assessments.

For grains still near the edge but not exhibiting HBS, although they initially possess suitable temperatures for recrystallization, their lower burn-up leads to a relatively lower accumulated dislocation density, a considerable portion of which is reduced by subgrain formation. In the analysis of radial positions slightly greater than 0.5R, experimental observations suggest that this location corresponds with specific temperature conditions: Gerczak et al. [26] observed this at 0.63R (about 64 GWd/tU), Cappia et al. [31] at 0.6R (about 76 GWd/tU), and Zacharie-Aubrun I et al. [33] indicated it is just above 0.5R (about 61 GWd/tU). At this position, the temperature has departed from the optimal conditions for recrystallization. Additionally, as the temperature rises, dislocation recovery is enhanced, reducing dislocation density and thus diminishing the occurrence of recrystallization. On the other hand, with a relatively low temperature compared to higher temperatures where fission gas becomes quite mobile, the volume proportion of pores is limited, leading to a minimal pinning effect on dislocations. This keeps the system energy at a lower level, rendering dislocation rearrangement insufficient to drive subgrain formation compared to the central regions, resulting in most grains still exhibiting original grain morphology.

3.2. Existing Research Supports Hybrid Mechanism Control

Experimental characterization results support this hypothesis to some extent. Gerczak et al. [26] noted that, as shown in Figure 3, near the center of the fuel, only the original grains and newly formed subgrains with low-angle boundaries (typically associated with the polygonization mechanism) are observed. At a relative radius of 0.63, there is a noticeable reduction in subgrain formation, a relative decrease in porosity, and the appearance of new grains with high-angle boundaries (commonly considered to be formed through recrystallization). Additionally, recrystallized grains appear at the boundary points between the original grains and subgrains, further corroborating the aforementioned nucleation perspective. As one moves outward to what is generally recognized as the thickness of the HBS, approximately 0.01R from the edge, there is an increase in recrystallized grains and subgrain formations, along with a rise in porosity distribution.

EPMA characterization results [26] further support this perspective, revealing a clear threshold boundary for grains at the edge of the fuel. When the position's energy exceeds a certain threshold (reflecting the combined effects of the aforementioned factors), there is a rapid increase in high-angle boundaries and a swift decrease in low-angle boundaries. This phenomenon may suggest that exceeding an energy threshold triggers the rapid onset of the recrystallization process.

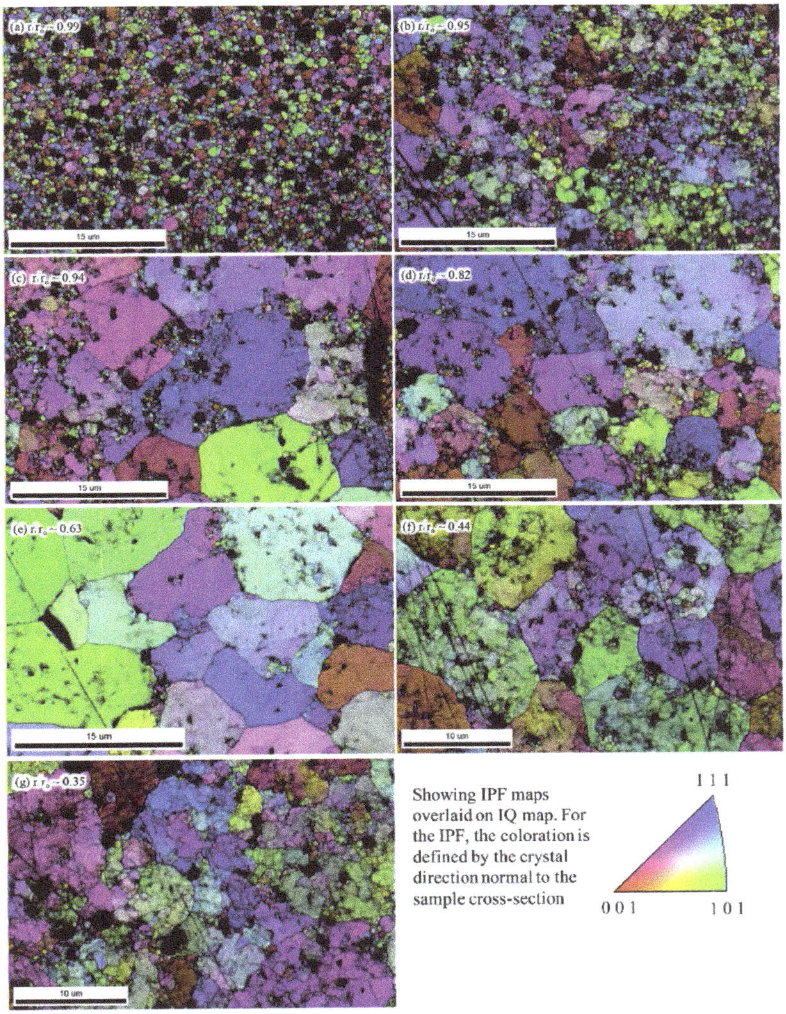

Figure 3. Locations of select EBSD scans moving from the fuel-cladding interface (r/r_0~0.99) toward the fuel center r/r_0~0.35, (**a**–**g**) [26].

A comparative observation of the SEM images shown in Figure 4, as provided by Xiao et al. [29], clearly shows the sequential appearance of low-angle boundaries with increasing burn-up, where low-angle and high-angle boundaries coexist, and high-angle boundaries are densely packed around the bubbles. This structural relationship with burn-up supports the hypothesis of a hybrid mechanism.

Veshchunov [48] referenced the experimental results of Une et al. [11], which demonstrated that under relatively low burn-up conditions of 44 GWd/tU at a temperature of 1000 K, low-angle boundaries were present. In contrast, at a higher burn-up of 83 GWd/tU, high-angle boundaries were observed. This also suggests the possibility of a threshold that serves as a boundary for a hybrid control mechanism.

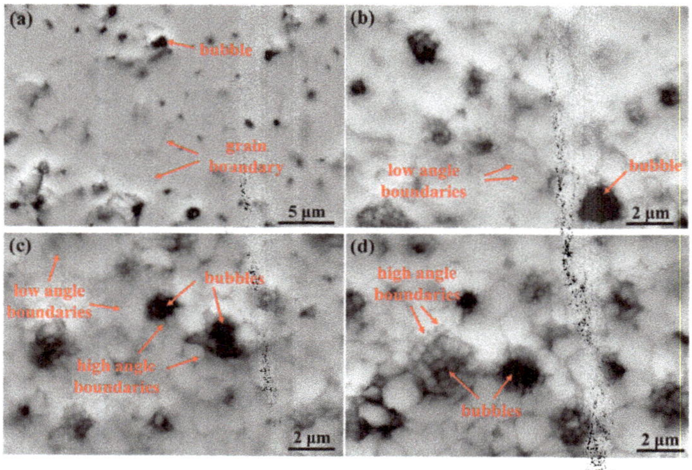

Figure 4. High magnification SEM micrographs illustrating the HBS with local burn-up of (**a**) 48 GWd/tU; (**b**) 87 GWd/tU; (**c**) 105 GWd/tU; and (**d**) 120 GWd/tU [29].

From a simulation perspective, there is a clear inclination toward exploring beyond a single mechanism. Jonnet [64] indicated that the future development of his polygonization-based model involves coupling it with Rest's recrystallization model, as mentioned above, implying that polygonization acts as a precursor to the recrystallization process. D. Pizzocri et al. [50] also noted that recrystallization and polygonization mechanisms are essentially related. Abdoelatef's model [53] treats grain refinement as a phase transformation without attempting to distinguish between recrystallization and polygonization principles. Although Baron et al. [69] considered polygonization to be the dominant mechanism, their energy assessments also partially support the hybrid mechanism viewpoint.

The experimental and simulation studies listed above indicate that a single control mechanism cannot fully explain the observed phenomena, indirectly supporting the proposal of a hybrid control mechanism.

4. Conclusions and Perspectives

4.1. Summary of Hybrid Mechanisms in HBS Formation

As described in the previous sections, it summarizes experimental observations related to the potential formation of HBS, the establishment of models based on these experiments, and the perspectives of various scholars. Starting from the two potential mechanisms for HBS formation—polygonization and recrystallization—and based on observations from experimental and simulation studies, it is proposed that dislocation evolution is a common feature between the two. It is further suggested that these mechanisms manifest differently under varying external conditions, reflecting distinct energy-based behaviors. Based on the proposed viewpoints, corresponding supporting experimental and simulation studies are listed, which, to some extent, substantiate the reasonableness of this hypothesis.

Regarding the hybrid control mechanism, it offers a better interpretation of the complete process from the emergence of low-angle boundary subgrains to the coexistence of recrystallized grains with high-angle boundaries and subgrains with low-angle boundaries, culminating in the significant appearance of pores primarily filled with recrystallized grains characterized by high-angle boundaries. This provides a theoretical foundation, as discussed in Section 3.2, for developing models to understand the microstructural evolution induced by changes in burn-up [26,29]. It also suggests the possibility of fully considering the impact of the dynamic microstructural evolution, which continuously changes with burn-up, on studies of fission gas behavior and nuclear fuels' performance degradation.

Recent simulation studies have generally focused on simulating the formation of recrystallization cores in the HBS region and the subsequent recrystallization processes, with little attention given to the formation of subgrains before the formation of recrystallization cores. Experimental observations [26] indicate that the boundary locations of these substructures also serve as nucleation sites for recrystallized grains, and pore accumulation at these sites suggests a trapping effect for fission gases, which could further induce recrystallization processes. The lack of theoretical elucidation and model establishment regarding grain size variations at different locations in PWR fuel rods may hinder the understanding of fuel behavior under high burn-up conditions and impede efforts to push fuel toward higher burn-ups.

4.2. Recommendations for Future Modeling and Experimental Research

Regarding recommendations for modeling work, first, it is essential to model the complete grain evolution process across the entire fuel cross-section, clarifying the conditions under which subgrains will form. For instance, the attempts made by Baranov et al. [70] to analyze this phenomenon using dislocation dynamics provide insights for modeling the polygonization mechanism that leads to subgrain formation. Based on the aforementioned hybrid mechanism control, a reasonable set of energy thresholds should be designed to vary with burn-up and temperature, allowing these thresholds to adequately describe both the dominant mechanisms suggested by most current experimental data and the transition regions, thus clarifying and refining the energy-related hybrid mechanism theory.

Secondly, it is crucial to consider the behavior of fission gases directly impacted by grain size evolution. For commonly used diffusion equations, grain size often determines diffusion distance, typically modeled using constant or step-function forms for normal micron-sized grains and HBS nano-grains, as well as exponential fitting or mechanistic grain growth models [34,71,72]. There is a lack of continuous and comprehensive consideration in grain refinement modeling. Additionally, since dislocation effects have been incorporated into the research of Rest et al. [34,72], it is worth investigating whether the subgrains formed by dislocation wall rotation may have a unique impact on fission gas behavior, especially since such trapping effects have been observed experimentally. Restani [73] also noted that the pressure and release conditions of the HBS, regardless of whether they are below the threshold release temperature, need to be considered separately. It is also important to address how various behaviors transform within different regions, such as the transition from intragranular bubbles to grain boundary bubbles, and how to identify internal regions, HBS transition regions, and HBS regions. Adjusting the corresponding parameters based on these distinctions will be crucial for analyzing fission gas behavior in the context of higher burn-ups.

Finally, the implications of grain evolution and the various impacts of the HBS should be integrated into analytical programs. The presence of grain boundaries significantly reduces the thermal conductivity of fuel, but the impact on subgrains has not been adequately addressed, nor has the evolution of grain size been included in commonly used thermal conductivity calculations. Currently, there are some attempts to account for the effects of HBS on thermal conductivity. Xian-Ming Bai et al. [65] conducted an analytical model on the effects of different Xe concentrations, bubble porosities, and grain sizes on thermal conductivity, aiming to evaluate how HBS impacts the thermal conductivity of UO_2. The larger porosity characteristic of HBS significantly reduces thermal conductivity and requires consideration of the effects during the depletion process. Gado et al. [69] considered the local changes caused by the formation of HBS in the FUROM code: the local porosity increases, which results in a decrease in the UO_2 density. Young's modulus is also notably affected by HBS, as Terrani et al. [9] have established a relationship between Young's modulus and burn-up. Furthermore, beyond studying the influencing parameters, future research may explore modeling from the perspective of thermal-mechanical coupling or irradiation-thermal-mechanical coupling, with an emphasis on uniformization. Kulacsy [68] developed a model based on the assumption that during irradiation, the pores reach dislocation punching overpressure, and the increased overpressure due to rising temperature leads to fuel fragmentation. This model explains the fracturing of HBS in the fuel pellet during an LOCA and successfully simulates the transient release of fission gases and the depletion of gases in the matrix. Capps et al. [66] suggest that HBS exhibits lower yield stress compared to fresh fuel, and the model calculations indicate that the fracturing of high burn-up zone fuel is not caused by thermal stress under LOCA conditions. Currently, programs such as MFPR [48], TRANSURANUS [50], BISON [74], and DIONISIO [75] have partially considered the influence of HBS. In summary, how to effectively incorporate these macroscopic phenomena observed in experiments into fuel performance analysis programs will be a critical research area for high burn-up fuels under operational or accident conditions.

For recommendations regarding experimental work, with the continuous advancement of experimental analysis techniques, it may be necessary to conduct further experimental studies at higher burn-up points. For example, Veshchunov [48] noted that the burn-up primarily characterized by polygonization (44 GWd/tU) is lower than the burn-up required for observing recrystallization (83 GWd/tU). Additionally, such observations are generally limited by sample numbers, lacking a comprehensive characterization of microstructural features across different burn-ups or, more broadly, across the entire fuel cross-section (reflecting varying burn-ups and temperatures). This limitation somewhat restricts experimental observations regarding the formation mechanisms of HBS.

Moreover, with the development of auxiliary tools and interdisciplinary approaches, the formation of HBS can be explored through new methods. For example, research by Kinoshita et al. [17] indicates that certain chemical elements play a role, while Soba et al. [76] and Frazer [77] et al. argue that elemental migration and the formation of precipitates have a significant impact. The method [67] of using MATLAB tools to process SEM images to restore bubble morphology, quantity distribution, and interconnection can deepen our understanding of the effect of bubbles on HBS formation. The FIB-SEM 3D examination [78] can also provide similar insights. These new insights and tools from various fields will provide fresh perspectives on the formation of HBS. Existing studies suggest that HBS result from the local minimization of internal energy, driven by the need to adapt to the harsh irradiation–thermal–mechanical conditions within the reactors. This structure may not solely bring adverse effects during fuel operation. For instance, Spino et al. [79] suggest that nano-crystalline UO_2 could serve as a potential high-performance fuel with gas retention capabilities and resistance to particulate cladding interactions. Its advantages regarding oxidation resistance have also been observed in studies by Baranov et al. [70], warranting further exploration. Multiscale experimental observations [80] are also essential for understanding the formation mechanisms of HBS.

Finally, the formation mechanism of HBS remains an open question, lacking decisive conclusions. The viewpoints presented in this paper are merely inferences drawn from the current body of research. The formation of HBS and their implications are unavoidable topics in extending the lifespan of commercial reactor fuels towards higher burn-up expectations, transcending mere academic discourse. Utilizing increasingly mature experimental characterization techniques and multi-scale coupled analysis methods, combined with perspectives from physics, chemistry, and materials science, will provide a reasonable solution to the challenge of HBS formation.

Author Contributions: Writing—original draft preparation, Z.G.; methodology, D.Y. (Dong Yan); writing—review and editing, P.L. and D.Y. (Di Yun). All authors have read and agreed to the published version of the manuscript.

Funding: Financial support provided by the National Key Research and Development Program of China (No. 2022YFB1902401) and the National Natural Science Foundation of China (No. U23B2071) is acknowledged.

Data Availability Statement: The data presented in this study are available on request from the corresponding authors.

Conflicts of Interest: The authors declare no conflicts of interest.

References

1. Matzke, H. On the Rim Effect in High Burnup UO_2 LWR Fuels. *J. Nucl. Mater.* **1992**, *189*, 141–148. [CrossRef]
2. Baron, D.; Kinoshita, M.; Thevenin, P.; Largenton, R. Discussion About Hbs Transformation in High Burn-Up Fuels. *Nucl. Eng. Technol.* **2009**, *41*, 199–214. [CrossRef]
3. Rest, J.; Hofman, G.L. Irradiation-Induced Recrystallization of Cellular Dislocation Networks in Uranium-Molybdenum Alloys. *MRS Online Proc. Libr. OPL* **2000**, *650*, R1.7. [CrossRef]
4. Ray, I.L.F. Observation of a High Burnup Rim-Type Structure in an Advanced Plutonium–Uranium Carbide Fuel. *J. Nucl. Mater.* **1997**, *250*, 242–243. [CrossRef]
5. Rondinella, V.V. The High Burn-Up Structure in Nuclear Fuel. *Mater. Today* **2010**, *13*, 24–32. [CrossRef]
6. Barney, W.K.; Wemple, B.D. *Metallography of Irradiated UO_2-Containing Fuel Elements*; Knolls Atomic Power Laboratory, General Electric Company: Niskayuna, NY, USA, 1958.
7. Bleiberg, M.L.; Berman, R.M.; Lustman, B. Effects of High Burn-Up on Oxide Ceramic Fuels. In Proceedings of the Radiation Damage in Reactor Materials. Part of the Proceedings of the Symposium on Radiation Damage in Solids and Reactor Materials, Venice, Italy, 7–11 May 1963.
8. Walker, C.T.; Staicu, D.; Sheindlin, M.; Papaioannou, D.; Goll, W.; Sontheimer, F. On the Thermal Conductivity of UO_2 Nuclear Fuel at a High Burn-Up of Around 100MWd/kgHM. *J. Nucl. Mater.* **2006**, *350*, 19–39. [CrossRef]
9. Terrani, K.A.; Balooch, M.; Burns, J.R.; Smith, Q.B. Young's Modulus Evaluation of High Burnup Structure in UO_2 with Nanometer Resolution. *J. Nucl. Mater.* **2018**, *508*, 33–39. [CrossRef]
10. Nogita, K.; Une, K. Irradiation-Induced Recrystallization in High Burnup UO_2 Fuel. *J. Nucl. Mater.* **1995**, *226*, 302–310. [CrossRef]
11. Nogita, K.; Une, K. Radiation-Induced Microstructural Change in High Burnup UO_2 Fuel Pellets. *Nucl. Instrum. Methods Phys. Res. Sect. B Beam Interact. Mater. At.* **1994**, *91*, 301–306. [CrossRef]
12. Une, K.; Hirai, M.; Nogita, K.; Hosokawa, T.; Suzawa, Y.; Shimizu, S.; Etoh, Y. Rim Structure Formation and High Burnup Fuel Behavior of Large-Grained UO_2 Fuels. *J. Nucl. Mater.* **2000**, *278*, 54–63. [CrossRef]
13. Spino, J.; Papaioannou, D. Lattice Parameter Changes Associated with the Rim-Structure Formation in High Burn-Up UO_2 Fuels by Micro X-Ray Di Raction. *J. Nucl. Mater.* **2000**, *281*, 146–162. [CrossRef]
14. Noirot, J.; Desgranges, L.; Lamontagne, J. Detailed Characterisations of High Burn-Up Structures in Oxide Fuels. *J. Nucl. Mater.* **2008**, *372*, 318–339. [CrossRef]
15. Noirot, J.; Aubrun, I.; Desgranges, L.; Hanifi, K.; Lamontagne, J.; Pasquet, B.; Valot, C.; Blanpain, P.; Cognon, H. High Burnup Changes in UO_2 Fuels Irradiated up to 83 GWd/t in M5® Claddings. *Nucl. Eng. Technol.* **2009**, *41*, 1–11. [CrossRef]
16. Noirot, J.; Zacharie-Aubrun, I.; Blay, T. Focused Ion Beam–Scanning Electron Microscope Examination of High Burn-Up UO_2 in the Center of a Pellet. *Nucl. Eng. Technol.* **2018**, *50*, 259–267. [CrossRef]
17. Sonoda, T.; Kinoshita, M.; Ray, I.L.F.; Wiss, T.; Thiele, H.; Pellottiero, D.; Rondinella, V.V.; Matzke, H. Transmission Electron Microscopy Observation on Irradiation-Induced Microstructural Evolution in High Burn-Up UO_2 Disk Fuel. *Nucl. Instrum. Methods Phys. Res. Sect. B Beam Interact. Mater. At.* **2002**, *191*, 622–628. [CrossRef]

18. Sonoda, T.; Kinoshita, M.; Ishikawa, N.; Sataka, M.; Iwase, A.; Yasunaga, K. Clarification of High Density Electronic Excitation Effects on the Microstructural Evolution in UO_2. *Nucl. Instrum. Methods Phys. Res. Sect. B Beam Interact. Mater. At.* **2010**, *268*, 3277–3281. [CrossRef]
19. Kinoshita, M.; Yasunaga, K.; Sonoda, T.; Iwase, A.; Ishikawa, N.; Sataka, M.; Yasuda, K.; Matsumura, S.; Geng, H.Y.; Ichinomiya, T.; et al. Recovery and Restructuring Induced by Fission Energy Ions in High Burnup Nuclear Fuel. *Nucl. Instrum. Methods Phys. Res. Sect. B Beam Interact. Mater. At.* **2009**, *267*, 960–963. [CrossRef]
20. Kinoshita, M.; Geng, H.Y.; Chen, Y.; Kaneta, Y.; Iwaswa, M.; Ohnuma, T.; Sonoda, T.; Yasugan, K.; Matsumura, S.; Yasuda, K. Study of Irradiation Induced Restructuring of High Burnup Fuel–The New Cross-over Project (NXO) to Study RIM-Structure Formation. In Proceedings of the International Topical Meeting on LWR Fuel Performance, TOPFUEL 2006, Salamanca, Spain, 22–26 October 2006; p. 248.
21. Geng, H.Y.; Chen, Y.; Kaneta, Y.; Kinoshita, M. Structural Behavior of Uranium Dioxide under Pressure by LSDA+ U Calculations. *Phys. Rev. B Condensed Matter Mater. Phys.* **2007**, *75*, 054111. [CrossRef]
22. Voter, A.F. Temperature-Accelerated Dynamics for Simulation of Infrequent Events. *J. Chem. Phys.* **2000**, *112*, 9599–9606.
23. Ichinomiya, T.; Uberuaga, B.P.; Sickafus, K.E.; Nishiura, Y.; Itakura, M.; Chen, Y.; Kaneta, Y.; Kinoshita, M. Temperature Accelerated Dynamics Study of Migration Process of Oxygen Defects in UO_2. *J. Nucl. Mater.* **2009**, *384*, 315–321. [CrossRef]
24. Miao, Y.; Yao, T.; Lian, J.; Zhu, S.; Bhattacharya, S.; Oaks, A.; Yacout, A.M.; Mo, K. Nano-Crystallization Induced by High-Energy Heavy Ion Irradiation in UO_2. *Scr. Mater.* **2018**, *155*, 169–174. [CrossRef]
25. Baranov, V.G.; Lunev, A.V.; Reutov, V.F.; Tenishev, A.V.; Isaenkova, M.G.; Khlunov, A.V. An Attempt to Reproduce High Burn-Up Structure by Ion Irradiation of SIMFUEL. *J. Nucl. Mater.* **2014**, *452*, 147–157. [CrossRef]
26. Gerczak, T.J.; Parish, C.M.; Edmondson, P.D.; Baldwin, C.A.; Terrani, K.A. Restructuring in High Burnup UO_2 Studied Using Modern Electron Microscopy. *J. Nucl. Mater.* **2018**, *509*, 245–259. [CrossRef]
27. Cappia, F.; Cullison, M.; Chen, T.; Kombaiah, B.; Bawane, K.; Teng, F.; Madden, J.; Perez, E.; Yao, T.; Lei, P.; et al. Grain Subdivision and Structural Modifications by High-Energy Heavy Ions in UO_2 with Different Initial Grain Size. *Nucl. Instrum. Methods Phys. Res. Sect. B Beam Interact. Mater. At.* **2022**, *515*, 48–60. [CrossRef]
28. Hiernaut, J.-P.; Wiss, T.; Colle, J.-Y.; Thiele, H.; Walker, C.T.; Goll, W.; Konings, R.J.M. Fission Product Release and Microstructure Changes during Laboratory Annealing of a Very High Burn-Up Fuel Specimen. *J. Nucl. Mater.* **2008**, *377*, 313–324. [CrossRef]
29. Xiao, H.; Long, C.; Chen, H. The Formation Mechanisms of High Burnup Structure in UO_2 Fuel. *J. Nucl. Mater.* **2021**, *556*, 153151. [CrossRef]
30. Barker, M.A.; Adam, M.S.; Little, P.; Mills, R.; Morgan, S.; Woodman, F.E. Observations of High Burnup Structure in AGR Fuel. *Prog. Nucl. Energy* **2022**, *153*, 104433. [CrossRef]
31. Cappia, F.; Wright, K.; Frazer, D.; Bawane, K.; Kombaiah, B.; Williams, W.; Finkeldei, S.; Teng, F.; Giglio, J.; Cinbiz, M.N.; et al. Detailed Characterization of a PWR Fuel Rod at High Burnup in Support of LOCA Testing. *J. Nucl. Mater.* **2022**, *569*, 153881. [CrossRef]
32. Wiss, T.; Thiele, H.; Janssen, A.; Papaioannou, D.; Rondinella, V.V.; Konings, R.J.M. Recent Results of Microstructural Characterization of Irradiated Light Water Reactor Fuels Using Scanning and Transmission Electron Microscopy. *JOM* **2012**, *64*, 1390–1395. [CrossRef]
33. Zacharie-Aubrun, I.; Dowek, R.; Noirot, J.; Blay, T.; Cabié, M.; Dumont, M. Restructuring in High Burn-Up UO_2 Fuels: Experimental Characterization by Electron Backscattered Diffraction. *J. Appl. Phys.* **2022**, *132*, 195903. [CrossRef]
34. Rest, J. *GRASS-SST: A Comprehensive, Mechanistic Model for the Prediction of Fission-Gas Behavior in UO_2-Base Fuels During Steady-State and Transient Conditions*; Argonne National Laboratory (ANL): Argonne, IL, USA, 1978. [CrossRef]
35. Lemes, M.; Soba, A.; Denis, A. An Empirical Formulation to Describe the Evolution of the High Burnup Structure. *J. Nucl. Mater.* **2015**, *456*, 174–181. [CrossRef]
36. Spino, J.; Stalios, A.D.; Santa Cruz, H.; Baron, D. Stereological Evolution of the Rim Structure in PWR-Fuels at Prolonged Irradiation: Dependencies with Burn-Up and Temperature. *J. Nucl. Mater.* **2006**, *354*, 66–84. [CrossRef]
37. Romano, A.; Horvath, M.I.; Restani, R. Evolution of Porosity in the High-Burnup Fuel Structure. *J. Nucl. Mater.* **2007**, *361*, 62–68. [CrossRef]
38. Cappia, F.; Pizzocri, D.; Schubert, A.; Van Uffelen, P.; Paperini, G.; Pellottiero, D.; Macián-Juan, R.; Rondinella, V.V. Critical Assessment of the Pore Size Distribution in the Rim Region of High Burnup UO_2 Fuels. *J. Nucl. Mater.* **2016**, *480*, 138–149. [CrossRef]
39. Rest, J.; Hofman, G.L. Dynamics of Irradiation-Induced Grain Subdivision and Swelling in $U3Si_2$ and UO_2 Fuels. *J. Nucl. Mater.* **1994**, *210*, 187–202. [CrossRef]
40. Rest, J.; Hofman, G.L. Effect of Recrystallization in High-Burnup UO_2 on Gas Release During RIA-Type Transients. *J. Nucl. Mater.* **1995**, *223*, 192–195. [CrossRef]

41. Rest, J. A Model for the Influence of Microstructure, Precipitate Pinning and Fission Gas Behavior on Irradiation-Induced Recrystallization of Nuclear Fuels. *J. Nucl. Mater.* **2004**, *326*, 175–184. [CrossRef]
42. Rest, J. A Generalized Model for Radiation-Induced Amorphization and Crystallization of U_3Si and U_3Si_2 and Recrystallization of UO_2. *J. Nucl. Mater.* **1997**, *240*, 205–214. [CrossRef]
43. Rest, J. Application of a Mechanistic Model for Radiation-Induced Amorphization and Crystallization of Uranium Silicide to Recrystallization of UO_2. *J. Nucl. Mater.* **1997**, *248*, 180–184. [CrossRef]
44. Rest, J. Derivation of Analytical Expressions for the Network Dislocation Density, Change in Lattice Parameter, and for the Recrystallized Grain Size in Nuclear Fuels. *J. Nucl. Mater.* **2006**, *349*, 150–159. [CrossRef]
45. Rest, J.; Hofman, G.L. An Alternative Explanation for Evidence That Xenon Depletion, Pore Formation, and Grain Subdivision Begin at Different Local Burnupsq. *J. Nucl. Mater.* **2000**, *277*, 231–238. [CrossRef]
46. Rest, J. A Model for the Effect of the Progression of Irradiation-Induced Recrystallization from Initiation to Completion on Swelling of UO_2 and U–10Mo Nuclear Fuels. *J. Nucl. Mater.* **2005**, *346*, 226–232. [CrossRef]
47. Thomas, L.E.; Beyer, C.E.; Chariot, L.A. Microstructural Analysis of LWR Spent Fuels at High Burnup. *J. Nucl. Mater.* **1992**, *188*, 80–89. [CrossRef]
48. Veshchunov, M.S.; Shestak, V.E. Model for Evolution of Crystal Defects in UO_2 Under Irradiation up to High Burn-Ups. *J. Nucl. Mater.* **2009**, *384*, 12–18. [CrossRef]
49. Xiao, H.; Long, C.; Chen, H. Model for Evolution of Grain Size in the Rim Region of High Burnup UO_2 Fuel. *J. Nucl. Mater.* **2016**, *471*, 74–79. [CrossRef]
50. Pizzocri, D.; Cappia, F.; Luzzi, L.; Pastore, G.; Rondinella, V.V.; Van Uffelen, P. A Semi-Empirical Model for the Formation and Depletion of the High Burnup Structure in UO_2. *J. Nucl. Mater.* **2017**, *487*, 23–29. [CrossRef]
51. Ray, I.L.F.; Thiele, H.A.; Kinoshita, M. An Electron Microscopy Study of the RIM Structure of a UO_2 Fuel with a High Burnup of 7.9% FIMA. *J. Nucl. Mater.* **1997**, *245*, 115–123. [CrossRef]
52. Jiang, Y.; La, Y.; Liu, X.; Liu, W. Effect of Grain Size on Irradiation-Induced Bubble Evolution in High Burn-Up UO_2: A Phase-Field Study. *J. Nucl. Mater.* **2024**, *601*, 155312. [CrossRef]
53. Abdoelatef, M.G.; Badry, F.; Schwen, D.; Permann, C.; Zhang, Y.; Ahmed, K. Mesoscale Modeling of High Burn-Up Structure Formation and Evolution in UO_2. *JOM* **2019**, *71*, 4817–4828. [CrossRef]
54. Abdoelatef, M.G. Mesoscale Modeling of Grain Refinement in Solids. Master's Thesis, Texas A&M University, College Station, TX, USA, 2019.
55. Ahmed, K.; Bai, X.; Zhang, Y.; Biner, B. *High-Burnup-Structure (HBS): Model Development in MARMOT for HBS Formation and Stability Under Radiation and High Temperature*; Idaho National Lab. (INL): Idaho Falls, ID, USA, 2016; pp. INL/EXT--16-40046, 1364505.
56. Zhang, Y.; Schwen, D.; Permann, C.; Bai, X.; Ahmed, K. *Model Development in MARMOT for High Burn-Up Structure Formation*; Idaho National Lab. (INL): Idaho Falls, ID, USA, 2017; pp. INL/EXT--17-43415-Rev000, 1473610.
57. Cheniour, A.; Hilty, F.W.; Petrie, C.M.; Capps, N.A. Mesoscale Modeling of the Effects of Accelerated Burnup on UO_2 Microstructural Evolution. *J. Nucl. Mater.* **2024**, *595*, 155036. [CrossRef]
58. Permann, C.J.; Jokisaari, A.M.; Tonks, M.R.; Schwen, D.; Gaston, D.R.; Kong, F.; Hiromoto, R.; Martineau, R.C. Scalable Feature Tracking for Finite Element Meshes Demonstrated with a Novel Phase-Field Grain Subdivision Model. *Nucl. Technol.* **2021**, *207*, 885–904. [CrossRef]
59. Staker, M.R.; Holt, D.L. The Dislocation Cell Size and Dislocation Density in Copper Deformed at Temperatures between 25 and 700 C. *Acta Metall.* **1972**, *20*, 569–579. [CrossRef]
60. Jiang, Y.; Xin, Y.; Liu, W.; Sun, Z.; Chen, P.; Sun, D.; Zhou, M.; Liu, X.; Yun, D. Phase-Field Simulation of Radiation-Induced Bubble Evolution in Recrystallized U–Mo Alloy. *Nucl. Eng. Technol.* **2022**, *54*, 226–233. [CrossRef]
61. Ding, R.; Guo, Z.X. Coupled Quantitative Simulation of Microstructural Evolution and Plastic Flow during Dynamic Recrystallization. *Acta Mater.* **2001**, *49*, 3163–3175. [CrossRef]
62. Jiang, W.; Hu, T.; Aagesen, L.K.; Biswas, S.; Gamble, K.A. A Phase-Field Model of Quasi-Brittle Fracture for Pressurized Cracks: Application to UO_2 High-Burnup Microstructure Fragmentation. *Theor. Appl. Fract. Mech.* **2022**, *119*, 103348. [CrossRef]
63. Baranov, V.G.; Lunev, A.V.; Tenishev, A.V.; Khlunov, A.V. Interaction of Dislocations in UO_2 during High Burn-Up Structure Formation. *J. Nucl. Mater.* **2014**, *444*, 129–137. [CrossRef]
64. Jonnet, J. A Contribution to the Understanding of the High Burn-Up Structure Formation in Nuclear Fuels. Ph.D. Thesis, Institut National Polytechnique de Lorraine, Nancy, France, 2007.
65. Bai, X.-M.; Tonks, M.R.; Zhang, Y.; Hales, J.D. Multiscale Modeling of Thermal Conductivity of High Burnup Structures in UO_2 Fuels. *J. Nucl. Mater.* **2016**, *470*, 208–215. [CrossRef]
66. Capps, N.; Sweet, R.; Harp, J.; Petrie, C.M. High-Burnup Fuel Stress Analysis Prior to and during a LOCA Transient. *J. Nucl. Mater.* **2021**, *556*, 153194. [CrossRef]

67. McKinney, C.; Seibert, R.; Helmreich, G.; Aitkaliyeva, A.; Terrani, K. Three-Dimensional Bubble Reconstruction in High Burnup UO_2. *J. Nucl. Mater.* **2020**, *532*, 152053. [CrossRef]
68. Humphreys, F.J.; Hatherly, M. *Recrystallization and Related Annealing Phenomena*; Elsevier: Amsterdam, The Netherlands, 2012; ISBN 0-08-098388-X.
69. Gadó, J.; Griger, Á.; Kulacsy, K. The Fuel Behaviour Code FUROM and Its High Burn-up Simulation Capabilities. *Nucl. Eng. Des.* **2018**, *327*, 274–285. [CrossRef]
70. Jegou, C.; Gennisson, M.; Peuget, S.; Desgranges, L.; Guimbretière, G.; Magnin, M.; Talip, Z.; Simon, P. Raman Micro-Spectroscopy of UOX and MOX Spent Nuclear Fuel Characterization and Oxidation Resistance of the High Burn-Up Structure. *J. Nucl. Mater.* **2015**, *458*, 343–349. [CrossRef]
71. Pizzocri, D.; Barani, T.; Luzzi, L. SCIANTIX: A New Open Source Multi-Scale Code for Fission Gas Behaviour Modelling Designed for Nuclear Fuel Performance Codes. *J. Nucl. Mater.* **2020**, *532*, 152042. [CrossRef]
72. Rest, J.; Zawadzki, S.A. *FASTGRASS: A Mechanistic Model for the Prediction of Xe, I, Cs, Te, Ba, and Sr Release from Nuclear Fuel Under Normal and Severe-Accident Conditions*; Nuclear Regulatory Commission: Washington, DC, USA; Argonne National Lab. (ANL): Argonne, IL, USA, 1992; pp. NUREG/CR-5840, ANL--92/3, 7255051.
73. Restani, R.; Horvath, M.; Goll, W.; Bertsch, J.; Gavillet, D.; Hermann, A.; Martin, M.; Walker, C.T. On the Condition of UO_2 Nuclear Fuel Irradiated in a PWR to a Burn-Up in Excess of 110 MWd/kgHM. *J. Nucl. Mater.* **2016**, *481*, 88–100. [CrossRef]
74. Hirschhorn, J.; Greenquist, I.; Wysocki, A.; Capps, N. Multiphysics Analysis of Fuel Fragmentation, Relocation, and Dispersal Susceptibility–Part 2: High-Burnup Steady-State Operating and Fuel Performance Conditions. *Ann. Nucl. Energy* **2023**, *192*, 109952. [CrossRef]
75. Denis, A.; Soba, A. Simulation of Pellet-Cladding Thermomechanical Interaction and Fission Gas Release. *Nucl. Eng. Des.* **2003**, *223*, 211–229. [CrossRef]
76. Soba, A.; Lemes, M.; González, M.E.; Denis, A.; Romero, L. Simulation of the Behaviour of Nuclear Fuel Under High Burnup Conditions. *Ann. Nucl. Energy* **2014**, *70*, 147–156. [CrossRef]
77. Frazer, D.; Cappia, F.; Harp, J.M.; Medvedev, P.G.; McClellan, K.J.; Voit, S.L.; Giglio, J.; Jädernäs, D.; Hosemann, P. Post-Irradiation Characterization of a High Burnup Mixed Oxide Fuel Rod with Minor Actinides. *J. Nucl. Mater.* **2022**, *562*, 153545. [CrossRef]
78. Noirot, J.; Dowek, R.; Zacharie-Aubrun, I.; Blay, T.; Cabié, M.; Dumont, M. Restructuring in High Burn-Up Pressurized Water Reactor UO_2 Fuel Central Parts: Experimental 3D Characterization by Focused Ion Beam—Scanning Electron Microscopy. *J. Appl. Phys.* **2022**, *132*, 195902. [CrossRef]
79. Spino, J.; Santa Cruz, H.; Birtcher, R.; Ferrero, C.; Pieritz, R.; Fernández, A. Development at ITU towards a High-Performance, Gas-Retentive and PCI-Resistant Nanocrystalline UO_2-Fuel. In Proceedings of the International Workshop on Radiation Stability of Complex Microstructures, Poster Section, Santa Fe, NM, USA, 2–4 September 2008.
80. Degueldre, C.; Bertsch, J.; Martin, M. Post Irradiation Examination of Nuclear Fuel: Toward a Complete Analysis. *Prog. Nucl. Energy* **2016**, *92*, 242–253. [CrossRef]

nanomaterials

Article

Micro-Defects-Related Low Cycle Fatigue Mechanical Model of the Nuclear-Grade S30408 Stainless Steel

Huiping Liu [1,2], Mingkun Xiao [1,2], Jiannan Hao [3], Xinjie Ma [3], Ni Jiang [1,2], Qing Peng [3,4,5,*] and Chao Ye [1,2,*]

1 Institute of Clean Energy, Yangtze River Delta Research Institute, Northwestern Polytechnical University, Taicang 215400, China
2 School of Materials Science and Engineering, Northwestern Polytechnical University, Xi'an 710072, China
3 State Key Laboratory of Nonlinear Mechanics, Institute of Mechanics, Chinese Academy of Sciences, Beijing 100190, China
4 Center of Materials Science and Optoelectronics Engineering, University of Chinese Academy of Sciences, Beijing 100049, China
5 Guangdong Aerospace Research Academy, Guangzhou 511458, China
* Correspondence: pengqing@imech.ac.cn (Q.P.); yechao@nwpu.edu.cn (C.Y.)

Abstract: Continuous and interrupted low cycle fatigue tests were conducted on nuclear-grade S30408 stainless steel under different stress conditions at room temperature. Vickers hardness testing and microstructure characterization were performed on the fatigue samples with different fatigue states. The evolutionary mechanism of the microstructure defects in materials under fatigue cyclic loading was discussed. The traditional Basquin formula was used to predict the fatigue life of these fatigue samples. At the same time, a quantitative mechanical model related to the characteristic micro-defects parameter KAM and the Vickers hardness (H_v) was established for the S30408 stainless steel during the low cycle fatigue damage process, and the prediction accuracy of the Vickers hardness is greater than 90%, which is significant and useful for the fatigue life prediction of the 304 stainless steels used in nuclear systems and the safe operation of the reactors.

Keywords: S304 stainless steel; low cycle fatigue; mechanical model; characteristic defects

1. Introduction

Due to its excellent heat resistance, corrosion resistance, weldability, and mechanical properties, austenitic 304 stainless steel is widely used in engineering applications [1], such as in automobile parts, ship components, and nuclear power structural components [2–4]. In reactors, 304 austenitic stainless steels are normally employed in the core material [5], main pipeline [6], fuel cladding [7], reactor vessel [8], pressurized water reactor [9] and so on. During the service process, the weakening of the metallic materials caused by static-dynamic coupling loading results in fatigue failure, which seriously affects reactor operational safety. Therefore, the fatigue performance and predicted fatigue life of 304 stainless steel have always been important research topics for the long-term safe service of reactor components.

Since the mid-19th century, scholars have conducted extensive research in the field of fatigue damage. The nuclear pressure vessels and nuclear piping components in reactors are typically subjected to relatively high stresses and low failure cycles, which is referred to as low-cycle fatigue [10]. By considering the effect of pre-cycling damage history, a method used to estimate the residual cycles to failure for SUS304 stainless steels was proposed [11].

Furthermore, the application of a small punch test (SPT) was developed and used to evaluate the residual fatigue life of 304 stainless steel [12]. The influence of pressure, temperature, wall thickness, and stagnation time on the fatigue life of 304 stainless steel was investigated using the mechanical fatigue model (MFM) [13]. Moreover, the impact of high-density electrical pulses on the fatigue life of 304 stainless steel was also examined [14]. Additionally, several non-destructive techniques and models utilized to assess the remaining life of 304 stainless steels were developed and verified [15–18]. Currently, there are two types of widely used damage mechanics models applied to describe the progression of fatigue damage: The first is the Chaboche model (and its various simplified models) [19,20], which is based on the fatigue damage curve directly constructing the damage evolution function. This model is more complex as it has a greater number of parameters. The second is the Lemaitre model [21–24], which is derived from thermodynamic dissipation theory and is essentially a ductile damage model, where the damage is directly related to the plastic strain.

Overall, there has been little research on the correlation between macroscopic mechanical properties and the microscopic defects of the steel during the fatigue damage service of the 304 stainless steel material in the past few decades. Thus, in this work, commercially purchased nuclear-grade S30408 austenitic stainless steel was employed to perform low-cycle fatigue tests under a range of stress conditions. Samples exhibiting varying cycle numbers in the intermediate unbroken stage were selected for mechanical property testing and microstructural characterization. A quantitative correlation model associating the characteristic mechanical parameters with the characteristic microscopic damage defects was established, which can help researchers to predict the fatigue life of 304 stainless steel more accurately from the microscopic damage mechanism.

2. Experiments and Methods

S30408 stainless steel produced by Shanxi Taigang Stainless Steel Co., Ltd., Taiyuan, China, is used as the research material in this research, and its elemental compositions and mechanical properties are shown in Tables 1 and 2, respectively.

Table 1. Chemical composition of S30408 stainless steel (wt.%).

Element	C	Si	Mn	P	S	Cr	Ni	N
Mass Percent	0.04	0.47	1.04	0.030	0.001	18.02	8.04	0.03

Table 2. Mechanical properties of S30408 stainless steel.

Brand Number	Yield Strength $R_{p0.2}$ (MPa)	Tensile Strength R_m (MPa)	Elongation A5 (%)	Elongation A50 (%)
S30408	281	710	59.0	52

First, S30408 stainless steels were processed and polished into fatigue specimens, as shown in Figure 1. Subsequently, in accordance with the standard for axial force control in the fatigue testing of metallic materials" [25], low cycle fatigue experiments—with the constant stress amplitude and stress ratio R = 0.1—were conducted under four stress conditions: 635 MPa, 660 MPa, 700 MPa, and 730 MPa, respectively. In addition, in order to conduct research on the damage changes inside the material during the fatigue processes, according to a series of fatigue life ratios, interrupted low cycle fatigue tests were taken to prepare several fatigue samples with an unbroken state under two selected stress conditions.

For the unbroken fatigue samples acquired through interrupted fatigue testing, the stress area in the middle section of these fatigue samples was selected for cutting to obtain block samples with the thickness of about 5 mm. Next, SiC sandpapers, with particle sizes ranging from 500 mesh to 4000 mesh, were used to polish the surface of the samples until it reached a mirror-like state, which was then followed by nanoindentation testing. The Vickers hardness values of the fatigue samples were measured by a Vickers hardness tester. Ten indentations were performed for each sample, and the results were averaged over the ten measurements.

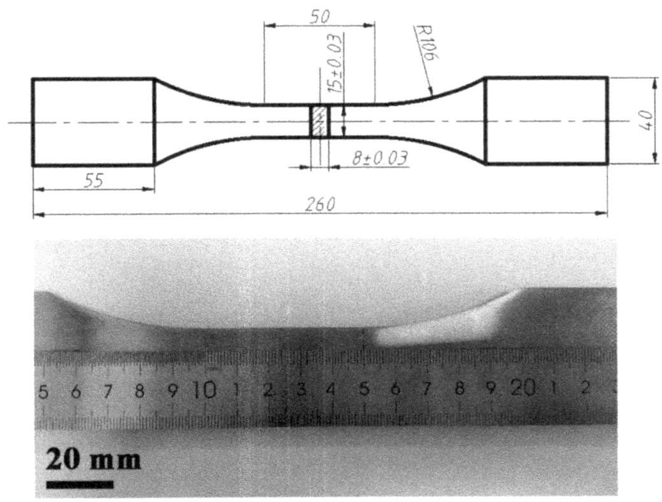

Figure 1. Morphology and size of the S30408 stainless steel fatigue specimens.

The fracture morphology and grain structure state of the fatigue samples were characterized via a scanning electron microscope (SEM, Zeiss Gemini 460) equipped with electron backscattered diffraction (EBSD) performance. A transmission electron microscope (TEM, FEI talos F200X), equipped with BF (bright field) and DF (dark field) STEM (scanning transmission electron microscopy) mode, was used to observe the microstructural defects inside the fatigue samples, and TEM samples with a thickness less than 100 nm were prepared using the focused ion beam (FIB, FEI Helios G4) lift-out technique.

3. Results

3.1. Mechanical Properties

The results of the fatigue tests are shown in Table 3. As the loading stress level gradually increases, the fatigue life of the S30408 stainless steel smooth specimen continues to decline. Based on the low cycle fatigue test parameters controlled by four stress levels and the data of the four low cycle fatigue tests, the maximum stress and fatigue life discrete points were fitted to obtain the stress–life curve, as shown in Figure 2. The quantitative relationship between the maximum stress (σ_{max}) and fatigue life (N_f) of the smooth S30408 specimens is as follows:

$$\sigma_{max} = 7354.57304\left(N_f\right)^{-0.2222} \tag{1}$$

The main micro-defects caused by fatigue are dislocations, which are always associated with the hardness of the material [26]. The statistical results of the variation in the Vickers hardness with different cycles of the unbroken fatigue samples under 660 MPa and 730

MPa stress fatigue tests are shown in Figure 3. It can clearly be seen from the figure that with the increase in cycles, the hardness of these fatigue samples shows an upward trend, and the increase is mainly concentrated after one cycle. In comparison, the hardness of the fatigue sample under 730 MPa stress is higher than that of the sample under 660 MPa stress at each cycle state.

Table 3. Results of the fatigue tests.

NO.	Max Stress (MPa)	Interrupted Fatigue State (Cycle)	Fatigue Life (Cycle)	Elongation (mm)
1	635	/	57,928	27
2	660	1, 3437, 6874, 13,749, 27498	54,997	33
3	700	/	38,981	39
4	730	1, 2054, 4108, 8217, 16,435	32,870	42

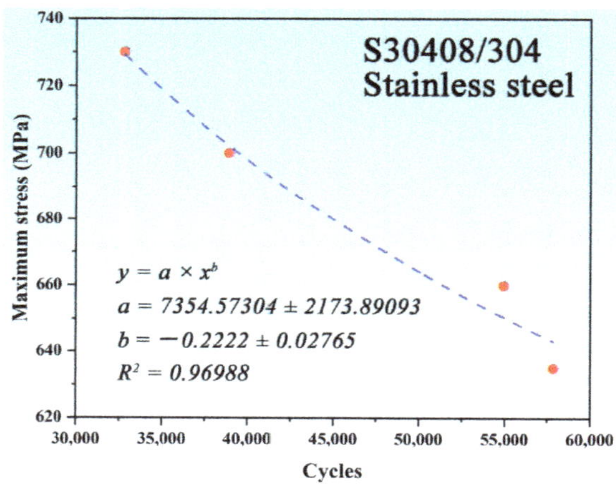

Figure 2. Stress–life curve of the S30408 stainless steel. The discrete red dots represent experimental data points, and the blue dashed line is the fitting curve.

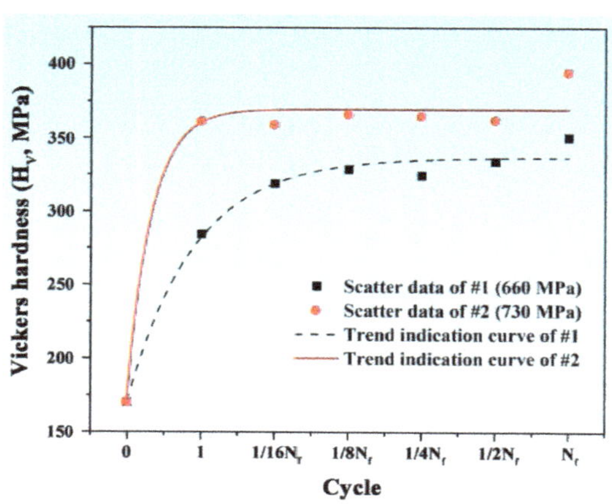

Figure 3. Statistical results of Vickers hardness of the S30408 stainless steel fatigue samples with different cycles.

3.2. Microstructure Characterization

The SEM images of the fracture morphologies of the S30408 fatigue samples are shown in Figure 4. No obvious crack source area can be observed on the fracture surface of the fatigue sample under 635 MPa stress (see Figure 4a), but there are obvious ductile dimples on the fracture surface (see Figure 4b). The fracture surface of the fatigue sample under 660 MPa stress clearly shows three different stages of the fatigue fracture process, namely, the slightly depressed fatigue source area on the surface of the fatigue sample shown in Figure 4d, which is generally located in areas with grooves, defects, or stress concentration on the surface and is a prerequisite for the generation of crack sources [27]. The crack source in Figure 4d is located at the center of the surface of the sample in the semi-circular area shown in Figure 4c, which is the crack propagation zone. The cross-section of this area is relatively flat, and the fatigue propagation is perpendicular to the stress direction, producing obvious fatigue arcs, also known as beach lines or shell lines [28,29]. Second, the outer side of the semicircle is referred to as the rapid fracture zone, also known as the instantaneous fracture zone, which is the area where fatigue cracks rapidly propagate to instantaneous fracture. The fracture surface has metal slip marks, and some fracture surfaces have radioactive stripes and shear lips in the instantaneous fracture zone. The fatigue samples under 700 MPa and 730 MPa stress—both fractured from the right angled edge region of the surface (see Figure 4e,g)—and the fracture surfaces show a ductile fracture dimple morphology and a serpentine slip morphology in the port extension region, which is consistent with common fatigue fracture morphology characteristics [30].

In addition, we also carried out EBSD comparative analysis on the original S30408 stainless steel samples and the fatigue fracture samples under 660 MPa stress conditions, as shown in Figure 5 below. Figure 5a,e are FSD diagrams (forescatter detector diagram and surface morphology synthesis); Figure 5b,f are IPF-X reverse pole diagrams (inverse pole figure map, grain orientation-related); Figure 5c,g are phase distribution diagrams (FCC phase and BCC phase distribution); Figure 5d,h are KAM diagrams (Kernel average misorientation is a method used to characterize the local dislocation supporting role angle in EBSD data analysis and is related to the distribution of stress, the dislocation grain boundary, and other defects, indicating the degree of plastic deformation). It can be seen that the shape, size, and orientation distribution of grains can be clearly distinguished from the FSD and IPF images of the original state sample. However, in the fatigue fracture sample, the grain boundaries and morphology distribution are difficult to distinguish. Comparing the phase distribution diagrams (Figure 5c,g), it can be seen that there is a significant increase in BCC phase after fatigue, which may be due to the strain-induced martensitic transformation process of the metastable phase in the original FCC austenite under fatigue-induced stress [31]. In addition, comparing the KAM results (Figure 5c,g) of the original and fractured samples, it can be seen that the fatigue process has severely disrupted the grain boundary orientation inside the sample, resulting in a large number of dislocations.

We conducted a statistical analysis on the KAM values of fatigue samples with different cycles under the 660 MPa stress condition, and the results are shown in Figure 6. Due to severe deformation of the internal grains of the material during the fatigue test, the analytical results of EBSD were not ideal. However, it can still be seen in Figure 6a that as the number of cycles increases, the overall stress inside the grains of the fatigue sample increases significantly, which is mainly due to the contribution of dislocations. And Figure 6b also clearly shows the trend of the KAM value increasing with the increase in cycle times, mainly concentrated after one cycle, which is consistent with the aforementioned trend of the hardness change.

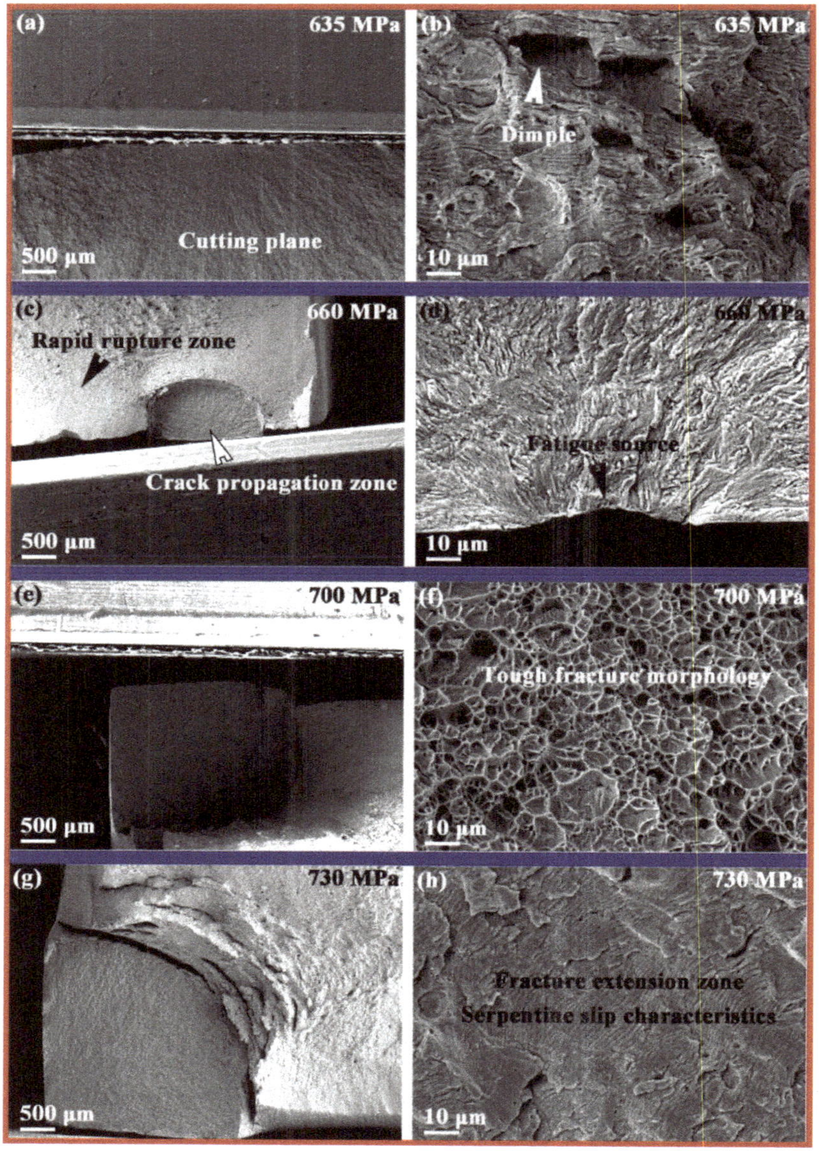

Figure 4. The overall fracture morphologies and local magnified morphologies of S30408 stainless steel fatigue samples under four different stress test conditions: (**a**,**b**) 635 MPa; (**c**,**d**) 660 MPa; (**e**,**f**) 700 MPa; (**g**,**h**) 730 MPa.

Figure 7 depicts the BF and DF STEM images of the microstructures of fatigue samples with different cycles under the 660MPa stress condition. Under bright-field conditions, dislocation structures exhibit black contrast; whereas under dark-field condition, dislocation structures exhibit white contrast. The interior of the original sample was mainly composed of strip-shaped dislocation lines, as shown in Figure 7a,b. After one cycle, a large number of densely packed dislocations were generated inside the material, and the dislocations began to entangle with each other, as shown in Figure 7c,d. A large number of obvious planar slip bands were generated inside the sample after $1/2N_f$ cycles, as shown in Figure 7e,f. After N_f cycles of fatigue fracture state, the internal planes of the sample formed wavy slip bands, as shown in Figure 7g,h.

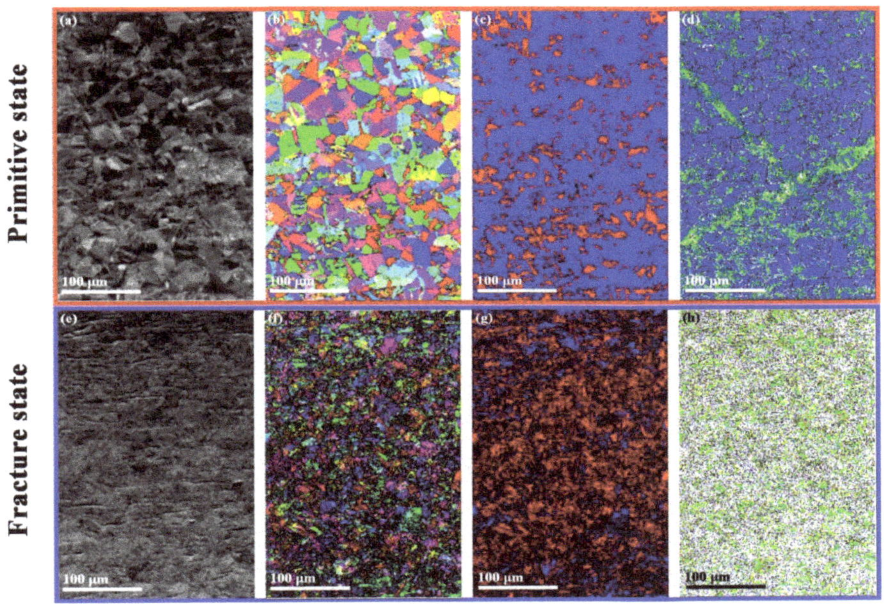

Figure 5. EBSD results of S30408 stainless steel samples under original state and 660 MPa stress fatigue fracture condition: (**a**,**e**) FSD plots; (**b**,**f**) IPF-X diagram; (**c**,**g**) phase distribution diagram; (**d**,**h**) KAM diagram. The different colors in the images represent the grains with different orientations and crystal structures.

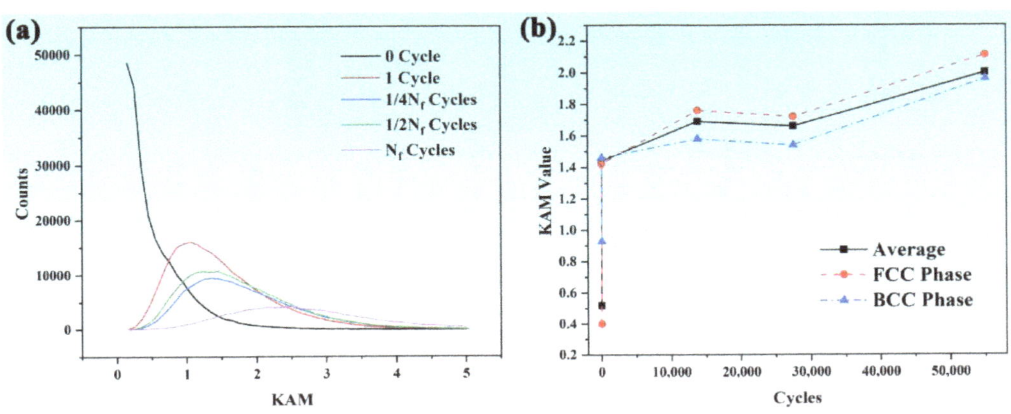

Figure 6. Defect statistics of the S30408 stainless steel fatigue samples under 660 MPa stress conditions: (**a**) KAM count statistics of fatigue fracture samples at different cycles; (**b**) trend chart of KAM variation with cycles.

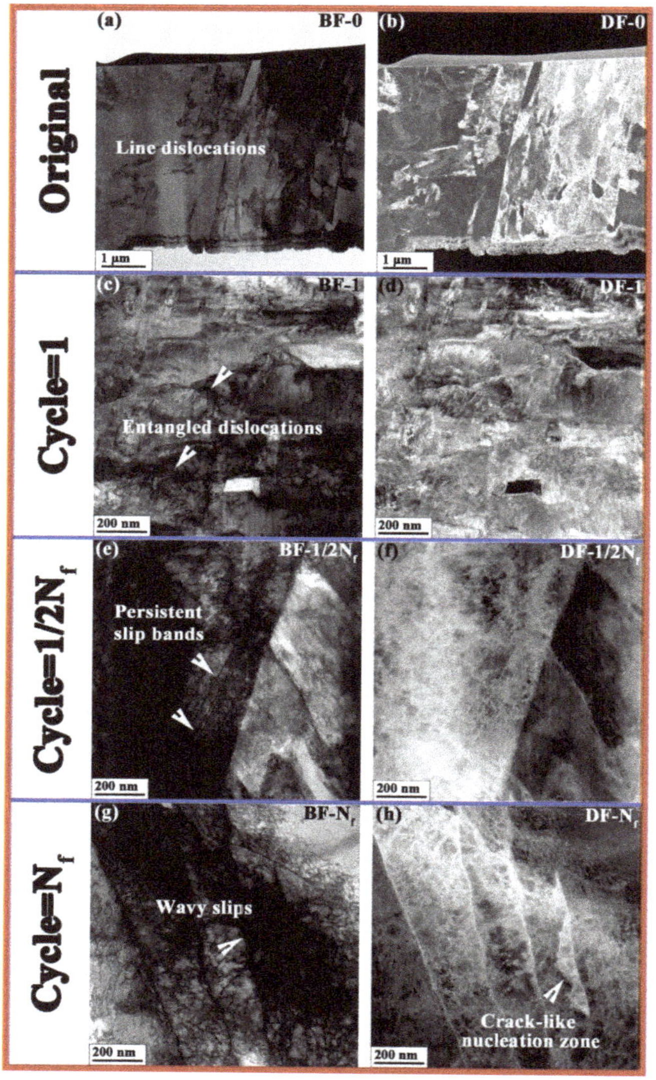

Figure 7. BF and DF STEM images of the S30408 stainless steel fatigue samples under 660 MPa stress condition.

4. Discussion

4.1. Microstructure

Currently, it is widely believed that macroscopic fatigue failures in metals can be attributed to deformation mechanisms occurring at the micro-nano scale within the internal grains of the material [32,33]. Figure 5 indicates that the S30408 sample underwent severe plastic deformation after fatigue fracture, and the grains size became much smaller as the grain breakage occurred, which is consistent with the KAM statistical results in Figure 6. In addition, the phase distribution diagram results illustrate that most of the face-centered cubic austenitic phases transformed into the body-centered cubic martensitic phases after fatigue damage, as seen in Figure 5c,g. It is generally believed that the austenite grain boundaries and dislocation concentration zones are usually cited as potential nucleation sites for α' martensite, and the crystallographic structures can be changed under plastic deformation [34,35]. During the process of fatigue damage, the increased nucleation sites accelerated the phase transformation rate, inducing the formation of nanotwins and fine martensite laths. Simultaneously, under the applied stress or strain, the Gibbs free energy of the austenite phase increased. When the energy difference between austenite and α' martensite reached the threshold for spontaneous transformation $\left(\Delta G_{RT}^{\gamma\to\alpha'} + U' \geq \Delta G_{Ms}^{\gamma\to\alpha'}\right)$ [36], the α' martensitic transformation began; the martensitic phase has a higher strength than the austenite phase, and the phase transformation made a certain contribution to the increase in hardness of the S30408 sample during the fatigue process. Additionally, the expansion of residual austenite volume and the plastic deformation induced by the compaction of surrounding soft phases during the phase transformation can also increase the material strength [31,32,37,38].

On the other hand, Figure 7a,e show that a large number of dislocations were generated and entangled after one cycle of fatigue loading, which was induced by plastic deformation [39–41]. And the persistent slip bands (PSBs) can clearly be observed in the sample after $1/2N_f$ cycles of fatigue loading, as shown in Figure 7c,g, which are referred to as the most consequential defect structures with regard to fatigue crack initiation [32]. With the continuation of the fatigue process, the PSBs accumulated and changed into wavy slip bands, a dislocation entanglement structure with larger dimensions, as shown in Figure 7d,h. After N_f cycles of fatigue loading, the PSBs almost occupied the entire sliding region, exceeding the stress threshold for crack formation and ultimately leading to fatigue failure of the material [42–45]. Overall, the entanglement of dislocations and the martensitic transformation are the main defect evolutions occurring inside the S30408 stainless steels during the fatigue loading processes, both of which can result in an increase in hardness [46,47].

4.2. Mechanical Models

According to the Basquin formula [48], during a constant stress amplitude fatigue test, there is a relationship between the stress amplitude and the number of load cycles at which failure occurs, as shown in the formula (2) below.

$$\sigma_a = \sigma_f'\left(2N_f\right)^b \tag{2}$$

where σ_a is the stress amplitude, σ_f' is the fatigue strength coefficient, N_f is the number of cycles at constant amplitude load fatigue fracture, and b is the fatigue strength index. Meanwhile, the four-point correlation method can be used to determine the fatigue strength index b [49].

$$b = -\left[0.083 + 0.166log\left(\frac{\sigma_f}{\sigma_b}\right)\right] \tag{3}$$

where σ_b is the tensile strength of the material during static tensile testing. Based on mechanical performance data in Table 2, using following formula (4) [50], which is taken as 1060 MPa, σ_b is taken as 710 MPa, and σ_f is the true fracture strength.

$$\sigma_f = \sigma_b + 350\text{MPa} \tag{4}$$

For convenience of calculation, it is generally believed that $\sigma_f{'} \approx \sigma_f$, so σ_f can be taken as 1060 MPa. Meanwhile, according to formula (3), b is calculated as -0.111892. Thus, the fatigue prediction model for S30408 stainless steel is as shown as below.

$$\sigma_a = 1060\left(2N_f\right)^{-0.111892} \quad (5)$$

According to the above fatigue model (5), the calculated data and experimental data are shown in Table 4. It can be seen that the theoretical calculation data are in good agreement with the experimental data. The mechanical performance prediction accuracy is greater than 90%, and the maximum error is 7.23%.

Table 4. The stress amplitude data calculated by the fatigue prediction model and obtained from the fatigue experiments.

No.	σ_{max} (MPa)	N_f (Cycle)	Cal. σ_{a1} (Mpa)	Exp. σ_{a2} (Mpa)	Error.$(\sigma_{a2}-\sigma_{a1})/\sigma_{a1}$ (%)
1	635	57928	287.5371032	285.75	−0.62%
2	660	54997	289.212464	297	2.69%
3	700	38981	300.5683836	315	4.80%
4	730	32870	306.3580274	328.5	7.23%

During the cyclic loading process of the fatigue load, the reciprocating motion of dislocations can lead to the self-organized growth of long and ordered dislocation structures, thereby causing the nucleation of microcracks in plastic metals. In EBSD characterization, the parameter *KAM* (Kernel average misorientation) can be used to quantitatively calculate the geometrically necessary dislocation density ρ_{GND}, reflecting the degree of homogenization of plastic deformation. Higher *KAM* values indicate greater plastic deformation or higher defect density in materials. Therefore, it was chosen as a quantitative parameter for the characteristic defects caused by fatigue damage in this research. The average KAM_{ave} is positively correlated with geometrically necessary dislocation density ρ_{GND}, satisfying the following relationship [51]:

$$\rho_{GND} = \frac{2KAM_{ave}}{\mu b} \quad (6)$$

where b is the dislocation Burger vector, and μ is the EBSD step size.

Different from the ideal laboratory environment, structural components are usually subjected to non-constant fatigue stress in practical engineering process. Therefore, in this study, a parameter, namely, simplified fatigue life $n = N/N_f$ (N is the current cycle number; N_f is the fracture cycle number), is defined to evaluate the current fatigue stage and remaining fatigue life of the fatigue samples with a type of percentage. The microstructure characterization results indicate that the dislocation density increased significantly at the early stage of fatigue and then linearly increased. Since the growth rate of dislocations remains basically unchanged when the loading frequency does not change significantly, the relationship between the linear fitting n and *KAM* is as follows:

$$KAM = KAM_0 + kn \quad (7)$$

where KAM_0 is the initial *KAM* value (not equal to 0), and $k = 9.37 \times 10^{-6}$ is the generation rate of dislocations with respect to the simplified fatigue life n. The linear fitting results are shown in Figure 8.

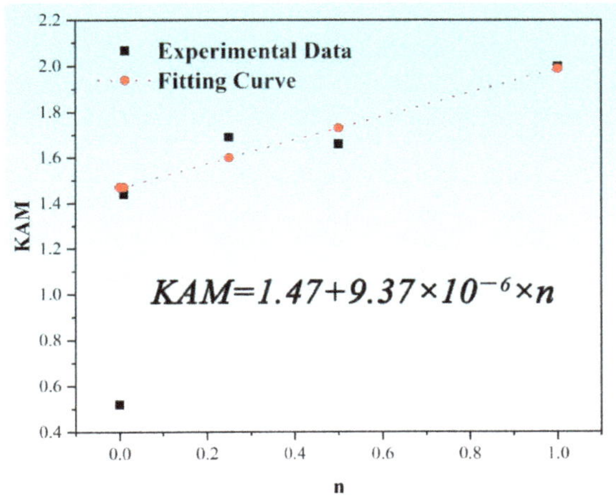

Figure 8. Linear fitting results of the simplified fatigue life n and KAM.

The relationship between flow stress σ and dislocation density ρ_{GND} is as follows [41,52,53]:

$$\sigma = AG\mu\sqrt{\rho_{GND}} \tag{8}$$

where G is the shear modulus, and A is a constant. The relationship between flow stress and hardness generally follows a linear relationship [54,55]. Therefore, by combining the above formulae, the relationship between Vickers hardness Hv and KAM can be fitted as in the following equation:

$$H_V = B\sqrt{KAM} + C \tag{9}$$

where B and C are both fitting parameters, and B is related to the elastic modulus and dislocation. When C is set to 0, the fitting relationship between Vickers hardness Hv and KAM is shown in Figure 9. The prediction accuracy of Vickers hardness is greater than 90%, and the maximum error is 4.81%.

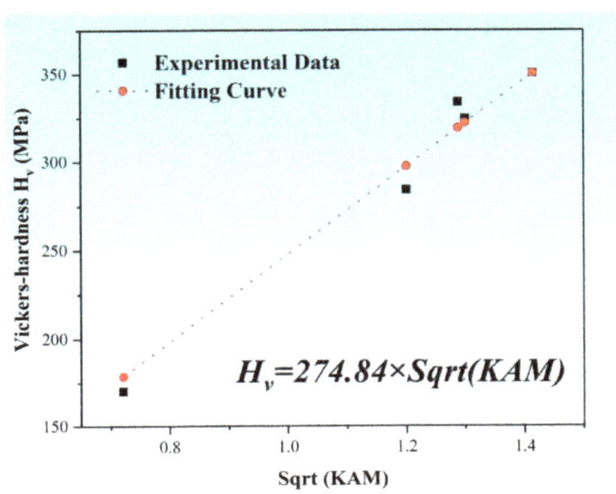

Figure 9. Linear fitting results of the H_v and KAM.

5. Conclusions

Low cycle fatigue tests of nuclear-grade S30408 stainless steels were performed in this work. And a new quantitative mechanical model was established to correlate the characteristic micro-defects induced by fatigue damage with the Vickers hardness. The specific conclusions are listed as follows:

(1) With the increase in fatigue cycle times, plastic deformation led to the continuous generation of dislocations. The reciprocating motion of dislocations resulted in the dislocation entanglement, ultimately leading to material failure due to formation of cracking.

(2) During the low cycle fatigue process, the KAM value and dislocation density increased accordingly, mainly concentrated after one cycle. And the metastable austenite phase underwent a phase transformation to a martensitic phase.

(3) The relationship between fatigue life (N_f) and loading stress (σ_a) of S30408 stainless steel satisfies the Basquin equation, and the prediction accuracy of stress amplitude is greater than 90%.

(4) A quantitative model linking fatigue micro characteristic defects and macroscopic mechanical properties was proposed. KAM and Vickers hardness are the characteristic defect parameters and mechanical property parameters, respectively. And the prediction accuracy of Vickers hardness in this model is greater than 90%.

Author Contributions: Conceptualization, C.Y.; methodology, Q.P. and C.Y.; software, X.M. and J.H.; formal analysis, H.L. and N.J.; investigation, M.X. and H.L.; writing—original draft preparation, H.L.; writing—review and editing, C.Y.; funding acquisition, C.Y. All authors have read and agreed to the published version of the manuscript.

Funding: This work was funded by the National Key R&D Program of China (Grant No. 2022YFB3707200).

Data Availability Statement: The data presented in this study are available on request from the corresponding authors.

Conflicts of Interest: The authors declare no conflicts of interest.

Abbreviations

FCC	Face-centered cubic
BCC	Body-centered cubic
KAM	Kernel average misorientation
$R_{p0.2}$	Yield strength (MPa)
R_m	Tensile strength (MPa)
A5, A50	Elongation (%)
R	Stress ratio
σ_{max}	Maximum stress (MPa)
N_f	Fatigue life
σ_a	Stress amplitude (MPa)
σ_f'	Fatigue strength coefficient (MPa)
b	Fatigue strength index
σ_f	True fracture strength (MPa)
σ_b	Tensile strength (MPa)
b_d	Dislocation burger vector
μ	EBSD step size
n	Simplified fatigue life
k	The generation rate of dislocations with respect to the simplified fatigue life n
σ	flow stress

G	Shear modulus
ρ_{GND}	Dislocation density
H_V	Vickers hardness (MPa)
A	A constant
B,C	Fitting parameters

References

1. Li, X.; Wei, D.; Zhang, S.; Qi, Y.; Sun, W.; Zhong, J.; Ning, G. Non-Negligible Corrosion from Vaporized Boric Acid on 304 Stainless Steel. *J. Nucl. Mater.* **2023**, *574*, 154188. [CrossRef]
2. Sun, C.; Zheng, S.; Wei, C.C.; Wu, Y.; Shao, L.; Yang, Y.; Hartwig, K.T.; Maloy, S.A.; Zinkle, S.J.; Allen, T.R.; et al. Superior Radiation-Resistant Nanoengineered Austenitic 304L Stainless Steel for Applications in Extreme Radiation Environments. *Sci. Rep.* **2015**, *5*, 7801. [CrossRef] [PubMed]
3. Haque, R.; Fairweather, H.; Olofinjana, A. Failure Analysis of Boat Rudder Fitting Fabricated from a Type 304 Stainless Steel. *J. Fail. Anal. Prev.* **2018**, *18*, 519–525. [CrossRef]
4. Li, W.; Chen, J.; Tatman, J.; Feng, Z.; Latour, M.; Miller, R.; Walters, L.; Rosseel, T.M.; Frederick, G. Weldability of Irradiated Stainless Steel 304 Materials Harvested from the National Research Universal (NRU) Reactor. *J. Nucl. Mater.* **2023**, *574*, 154193. [CrossRef]
5. Hoffman, A.; Arivu, M.; Wen, H.; Wu, Y. Advanced Characterization of Phase Stability Under Ion Irradiation of Utrafine-Grained and Nanocrystalline SS304L. *Microsc. Microanal.* **2020**, *26*, 402–403. [CrossRef]
6. Zangeneh, S.; Lashgari, H.R.; Sharifi, H.R. Fitness-for-Service Assessment and Failure Analysis of AISI 304 Demineralized-Water (DM) Pipeline Weld Crack. *Eng. Fail. Anal.* **2020**, *107*, 104210. [CrossRef]
7. Garner, F.A.; Makenas, B.J.; Chastain, S.A. Swelling and Creep Observed in AISI 304 Fuel Pin Cladding from Three MOX Fuel Assemblies Irradiated in EBR-II. *J. Nucl. Mater.* **2011**, *413*, 53–61. [CrossRef]
8. Yang, L.; Sun, K.; Peng, W.; Li, X.; Zhang, L. Effects of Grain Boundary Angles on Initial Deformation of 304 Austenitic Stainless Steel under Nanoindentation: A Molecular Dynamics Simulation. *Crystals* **2022**, *12*, 58. [CrossRef]
9. Wu, H.; Liu, X.; Xu, C.; Li, Y.; Yin, J.; Jin, X.; Jia, W.; Qian, W.; Wang, P.; Zhang, Y. Study on Corrosion Fatigue Behavior of 304L Austenite Stainless Steel in 325 °C High-Temperature Water Environment. *Metals* **2024**, *14*, 489. [CrossRef]
10. Vainionpää, A.; Seppänen, T.; Que, Z. Effects of Pressurized Water Reactor Environment and Cyclic Loading Parameters on the Low Cycle Fatigue Behavior of 304L Stainless Steel. *Int. J. Fatigue* **2024**, *182*, 108231. [CrossRef]
11. Chou, L.-H.; Chiou, Y.-C.; Huang, Y.-J. The Effects of Pre-Cycle Damage on Subsequent Material Behavior and Fatigue Resistance of SUS 304 Stainless Steel. *Mater. Sci. Eng. A* **2015**, *636*, 320–325. [CrossRef]
12. Tasdighi, E.; Nobakhti, H.; Soltani, N. Application of Small Punch Test in Predicting the Axial Fatigue Life of 304 Stainless Steel Sheets. *Exp. Tech.* **2016**, *40*, 1349–1357. [CrossRef]
13. Erinosho, T.O.; Li, P.; Truman, C.E.; Smith, D.J. Pressure, Temperature and Dwell Time Effects on Fatigue Life in 304 Stainless Steel Using a R5-Based Mechanistic Fatigue Model. *Procedia Eng.* **2016**, *160*, 191–198. [CrossRef]
14. Lesiuk, G.; Zimniak, Z.; Wiśniewski, W.; Correia, J.A.F.O. Fatigue Lifetime Improvement in AISI 304 Stainless Steel Due to High-Density Electropulsing. *Procedia Struct. Integr.* **2017**, *5*, 928–934. [CrossRef]
15. Mehdizadeh, M.; Khonsari, M.M. On the Application of Fracture Fatigue Entropy to Variable Frequency and Loading Amplitude. *Theor. Appl. Fract. Mech.* **2018**, *98*, 30–37. [CrossRef]
16. Mei, T.; Wang, Q.; Liu, M.; Jiang, Y.; Zou, T.; Cai, Y. The Low-Cycle Fatigue Behavior, Microstructure Evolution, and Life Prediction of SS304: Influence of Temperature. *Materials* **2023**, *16*, 6326. [CrossRef]
17. Ji, S.; Liu, C.; Li, Y.; Shi, S.; Chen, X. Effect of Torsional Pre-Strain on Low Cycle Fatigue Performance of 304 Stainless Steel. *Mater. Sci. Eng. A* **2019**, *746*, 50–57. [CrossRef]
18. Paventhan, R.; Lakshminarayanan, P.R.; Balasubramanian, V. Fatigue Behaviour of Friction Welded Medium Carbon Steel and Austenitic Stainless Steel Dissimilar Joints. *Mater. Des.* **2011**, *32*, 1888–1894. [CrossRef]
19. Chaboche, J.L.; Lesne, P.M. A Non-Linear Continuous Fatigue Damage Model. *Fatigue Fract. Eng. Mater. Struct.* **1988**, *11*, 1–17. [CrossRef]
20. Yang, X. A Continuous Low Cycle Fatigue Damage Model and Its Application in Engineering Materials. *Int. J. Fatigue* **1997**, *19*, 687–692. [CrossRef]
21. Lippmann, H. *A Course on Damage Mechanics*, 2nd ed.; Springer: Berlin/Heidelberg, Germany, 1996; ISBN 978-3-642-18255-6.
22. Kunc, R.; Prebil, I. Low-Cycle Fatigue Properties of Steel 42CrMo4. *Mater. Sci. Eng. A* **2003**, *345*, 278–285. [CrossRef]
23. Wang, T.-J. Unified CDM Model and Local Criterion for Ductile Fracture—II. Ductile Fracture Local Criterion Based on the CDM Model. *Eng. Fract. Mech.* **1992**, *42*, 185–193. [CrossRef]

24. Chandrakanthṭ, S.; Pandey, P.C. An Isotropic Damage Model For Ductile Material. *Eng. Fract. Mech.* **1995**, *50*, 457–465. [CrossRef]
25. *GB/T 3075-2008*; Metallic materials-Fatigue testing-Axial-force-controlled method. Standardization Administration of China: Beijing, China, 2008.
26. He, B.B.; Hu, B.; Yen, H.W.; Cheng, G.J.; Wang, Z.K.; Luo, H.W.; Huang, M.X. High Dislocation Density–Induced Large Ductility in Deformed and Partitioned Steels. *Science* **2017**, *357*, 1029–1032. [CrossRef] [PubMed]
27. Wang, M.; Shi, S.; Fineberg, J. Tensile Cracks Can Shatter Classical Speed Limits. *Science* **2023**, *381*, 415–419. [CrossRef] [PubMed]
28. Dejun, K. Effects of Micro Arc Oxidation on Fatigue Limits and Fracture Morphologies of 7475 High Strength Aluminum Alloy. *J. Alloys Compd.* **2015**, *650*, 393–398. [CrossRef]
29. Ni, H.; Zhang, S.; Lei, H. Experimental Study on Variable Amplitude Fatigue Performance of High-Strength Bolts in Steel Structure Flange Connections. *Buildings* **2024**, *14*, 3736. [CrossRef]
30. Milella, P.P. Morphological Aspects of Fatigue Crack Formation and Growth. In *Fatigue and Corrosion in Metals*; Springer Milan: Milan, Italy, 2013; pp. 73–108. ISBN 978-88-470-2335-2.
31. Das, A.; Tarafder, S.; Chakraborti, P.C. Estimation of Deformation Induced Martensite in Austenitic Stainless Steels. *Mater. Sci. Eng. A* **2011**, *529*, 9–20. [CrossRef]
32. Lavenstein, S.; Gu, Y.; Madisetti, D.; El-Awady, J.A. The Heterogeneity of Persistent Slip Band Nucleation and Evolution in Metals at the Micrometer Scale. *Science* **2020**, *370*, eabb2690. [CrossRef] [PubMed]
33. Ye, D.; Xu, Y.; Xiao, L.; Cha, H. Effects of Low-Cycle Fatigue on Static Mechanical Properties, Microstructures and Fracture Behavior of 304 Stainless Steel. *Mater. Sci. Eng. A* **2010**, *527*, 4092–4102. [CrossRef]
34. Ganesh Sundara Raman, S.; Padmanabhan, K.A. Influence of Martensite Formation and Grain Size on Room Temperature Low Cycle Fatigue Behaviour of AISI 304LN Austenitic Stainless Steel. *Mater. Sci. Technol.* **1994**, *10*, 614–620. [CrossRef]
35. Hahnenberger, F.; Smaga, M.; Eifler, D. Microstructural Investigation of the Fatigue Behavior and Phase Transformation in Metastable Austenitic Steels at Ambient and Lower Temperatures. *Int. J. Fatigue* **2014**, *69*, 36–48. [CrossRef]
36. Sohrabi, M.J.; Naghizadeh, M.; Mirzadeh, H. Deformation-Induced Martensite in Austenitic Stainless Steels: A Review. *Arch. Civ. Mech. Eng.* **2020**, *20*, 124. [CrossRef]
37. Yin, T.W. Controllable Selection of Martensitic Variant Enables Concurrent Enhancement of Strength and Ductility in a Low-Carbon Steel. *Int. J. Plast.* **2023**, *168*, 103704. [CrossRef]
38. Cao, B.; Iwamoto, T.; Bhattacharjee, P.P. An Experimental Study on Strain-Induced Martensitic Transformation Behavior in SUS304 Austenitic Stainless Steel during Higher Strain Rate Deformation by Continuous Evaluation of Relative Magnetic Permeability. *Mater. Sci. Eng. A* **2020**, *774*, 138927. [CrossRef]
39. Nabarro, F.R.N.; Hirth, J.P. *Dislocations in Solids*; Elsevier North Holland: Amsterdam, The Netherlands, 2007.
40. Hirth, J.P.; Lothe, J.; Mura, T. Theory of Dislocations (2nd Ed.). *J. Appl. Mech.* **1983**, *50*, 476–477. [CrossRef]
41. Taylor, G.I. The Mechanism of Plastic Deformation of Crystals. Part I.—Theoretical. *Proc. R. Soc. Lond. A.* **1934**, *145*, 362–387. [CrossRef]
42. Höppel, H.W.; Goik, P.; Krechel, C.; Göken, M. Ex and in Situ Investigations on the Role of Persistent Slip Bands and Grain Boundaries in Fatigue Crack Initiation. *J. Mater. Res.* **2017**, *32*, 4276–4286. [CrossRef]
43. Lavenstein, S.; El-Awady, J.A. Micro-Scale Fatigue Mechanisms in Metals: Insights Gained from Small-Scale Experiments and Discrete Dislocation Dynamics Simulations. *Curr. Opin. Solid State Mater. Sci.* **2019**, *23*, 100765. [CrossRef]
44. Straub, T.; Berwind, M.F.; Kennerknecht, T.; Lapusta, Y.; Eberl, C. Small-Scale Multiaxial Setup for Damage Detection Into the Very High Cycle Fatigue Regime. *Exp. Mech.* **2015**, *55*, 1285–1299. [CrossRef]
45. Sangid, M.D. The Physics of Fatigue Crack Initiation. *Int. J. Fatigue* **2013**, *57*, 58–72. [CrossRef]
46. Farahani, E.B.; Aragh, B.S.; Juhre, D. Interplay of Fracture and Martensite Transformation in Microstructures: A Coupled Problem. *Materials* **2022**, *15*, 6744. [CrossRef]
47. Tabata, T.; Fujita, H.; Hiraoka, M.-A.; Onishi, K. Dislocation Behaviour and the Formation of Persistent Slip Bands in Fatigued Copper Single Crystals Observed by High-Voltage Electron Microscopy. *Philos. Mag. A* **1983**, *47*, 841–857. [CrossRef]
48. Oh, B. The Exponential Law of Endurance Tests. *Am. Soc. Test. Mater. Proc.* **1910**, *10*, 625–630.
49. Morrow, J.D. *Fatigue Design Handbook Advances in Engineering*; Society of Automotive Engineers: Warrendale, PA, USA, 1968; pp. 21–29.
50. Li, S.M. *Mechanical Fatigue and Reliability Design*; Science Publishers: Beijing, China, 2006; pp. 113–114.
51. Yan, Z.; Wang, D.; He, X.; Wang, W.; Zhang, H.; Dong, P.; Li, C.; Li, Y.; Zhou, J.; Liu, Z.; et al. Deformation Behaviors and Cyclic Strength Assessment of AZ31B Magnesium Alloy Based on Steady Ratcheting Effect. *Mater. Sci. Eng. A* **2018**, *723*, 212–220. [CrossRef]
52. Basinski, Z.S. Forest Hardening in Face Centred Cubic Metals. *Scr. Metall.* **1974**, *8*, 1301–1307. [CrossRef]

53. Franciosi, P.; Berveiller, M.; Zaoui, A. Latent Hardening in Copper and Aluminium Single Crystals. *Acta Metall.* **1980**, *28*, 273–283. [CrossRef]
54. Zhu, Y.; Xiang, Y.; Schulz, K. The Role of Dislocation Pile-up in Flow Stress Determination and Strain Hardening. *Scr. Mater.* **2016**, *116*, 53–56. [CrossRef]
55. Rudnytskyj, A.; Varga, M.; Krenn, S.; Vorlaufer, G.; Leimhofer, J.; Jech, M.; Gachot, C. Investigating the Relationship of Hardness and Flow Stress in Metal Forming. *Int. J. Mech. Sci.* **2022**, *232*, 107571. [CrossRef]

nanomaterials

MDPI

Article

The Impact of Laser Irradiation on Thin ZrN Films Deposited by Pulsed DC Magnetron Sputtering

Ameena Nazneen, Penghui Lei * and Di Yun *

School of Nuclear Science and Technology, Xi'an Jiaotong University, Xi'an 710049, China; nazneen.ameena@stu.xjtu.edu.cn

* Correspondence: penghuilei@xjtu.edu.cn (P.L.); diyun1979@mail.xjtu.edu.cn (D.Y.)

Abstract: Transition metal nitrides have extensive applications, including magnetic storage devices, hardware resistance coatings, and low-temperature fuel cells. This study investigated the structural, electrical, and mechanical properties of thin zirconium nitride (ZrN) films by examining the effects of laser irradiation times. Thin ZrN films were deposited on glass substrates using pulsed DC magnetron sputtering and irradiated with a diode laser for 6 and 10 min. Characterization was performed using X-ray diffraction (XRD), scanning electron microscopy (SEM), high-resolution transmission electron microscopy (HRTEM), nanoindentation, and four-point probe techniques. Extended laser irradiation times resulted in increased numbers of peaks on XRD analysis, indicating enhanced crystalline behavior of thin ZrN film. SEM analysis revealed surface voids, while HRTEM showed nanostructured ZrN with uniform plane orientation. The electrical properties of the thin ZrN film improved with extended laser irradiation, as demonstrated by a reduction in sheet resistance from 0.43×10^9 Ω to 0.04×10^9 Ω. Additionally, nanoindentation tests revealed an increase in hardness, rising from 8.91 GPa to 9.36 GPa.

Keywords: zirconium nitride; sputtering; voids; thin film; laser

1. Introduction

In recent years, the demand for high-performance transition metal nitrides (TMNs) in the rock salt phase has been steadily increasing. These TMNs have found diverse applications, including magnetic storage devices, cutting tools, optical coatings, optoelectronic devices, resistance coating for generators, and low-temperature fuel cells. The exceptional properties of TMNs, derived from their mixed ionic, covalent, and metallic bonding, render them unique in amalgamating metallic and non-metallic traits. These properties encompass high melting points, luster, brittleness, high hardness, and high conductivity [1–3].

The field of hard-coating metal substrate applications has witnessed substantial growth recently [4]. Among these nitrides, zirconium nitride (ZrN) films have emerged as particularly interesting materials for a wide array of applications, including neutron reactors [5], aerospace components, cryogenic thermometers, diffusion barriers, decorative coatings, and hard coatings. ZrN films are favored for their exceptional corrosion and oxidation resistance [6], superior mechanical properties, lower resistivity, and distinctive warm golden color in comparison to titanium nitride (TiN) [7–10].

However, the deposition of ZrN films poses specific challenges when compared to materials like TiN and CrN due to zirconium's higher melting point, lower vapor pressure, and increased susceptibility to oxygen and carbon contamination [11]. The mono-nitrides of fourth-column transition metals exhibit interesting properties resulting from the combination of covalent and metallic bonding characteristics. Covalent crystalline properties contribute to high melting points, exceptional hardness, brittleness, and excellent thermal and chemical inertness, while the metallic attributes include electrical conductivity and metallic reflectance. These materials exhibit a gold-like appearance due to their high reflectance in the red end of the visible spectrum coupled with low reflectance in the ultraviolet region, thus presenting a unique optical signature [12–14]. Moreover, zirconium compounds are renowned for their chemical inertness, rendering them suitable for applications such as inert matrix fuel in nuclear reactors [9,15–19].

Physical vapor deposition (PVD) techniques, including RF sputtering, arc evaporation, and ion beam-assisted deposition, have been extensively employed for depositing transition metal nitride films. The properties of zirconium nitride (ZrN) films are significantly influenced by the deposition methods and associated processing parameters [11,20–22]. In magnetron sputtering, ion bombardment is widely recognized for its ability to modify film structure, enhancing film properties [23–26]. Pulsed DC magnetron sputtering has become a promising deposition technology, especially for minimizing arcing events caused by charge buildup on target surfaces. This reduction in arcing helps to maintain stable deposition conditions and prevent the degradation of film properties. The physical characteristics of ZrN films produced via pulsed DC magnetron sputtering are largely influenced by process parameters. For example, researchers studied how substrate temperature affects the tribological properties of thin ZrN films [27,28]. Laser irradiation enhances the surface mobility of condensed species, facilitating the formation of voids, increased crystallinity, and improved the material's hardness. Additionally, ZrO_2 can form in thin ZrN films as a result of zirconium reacting with oxygen, driven by oxygen diffusion [29–37]. Given that some physical properties of zirconium nitride films are closely related to their microstructure and nitrogen content, this study explored how laser irradiation affects thin ZrN films' microstructure and electrical and mechanical properties.

In this study, the pulsed DC (direct current) magnetron sputtering technique was chosen for depositing ZrN films onto glass substrates. ZrN coatings, known for their excellent mechanical, decorative, and wear-resistant properties, have been widely used for decades. However, in this work, we aimed to enhance the microstructural, mechanical, and electrical properties of the coatings by incorporating laser irradiation as a post-deposition treatment. Notably, to our knowledge, no prior research has reported the influence on ZrN films using laser surface treatment for 6 and 10 min with this specific laser technique. Our experimental findings suggest that laser irradiation can significantly improve the hardness, crystallinity, and electrical performance of these coatings. This approach has the potential to extend the functional range of thin ZrN films, offering new possibilities for high-performance applications. In this work, the physical, structural, morphological, electrical, and mechanical properties of the films were thoroughly investigated through X-ray diffraction (XRD), scanning electron microscopy (SEM), high-resolution transmission electron microscopy (HRTEM), four-point probe, and nanoindentation techniques.

2. Materials and Methods

Thin zirconium nitride (ZrN) films were deposited on glass substrates utilizing a pulsed DC magnetron sputtering system from AJA International, Hingham, MA, USA (AJA ATC 2200) within a vacuum chamber measuring 300 mm in diameter and 350 mm in height. Glass substrates were initially cleaned with a soap solution, then subjected to ultrasonic cleaning in acetone for 15 min, and finally dried with hot air before being placed in a deposition chamber. A high-purity metallic zirconium target (99.99%) with a 3-inch diameter was employed for sputtering. The deposition chamber was evacuated to a base pressure of 2×10^{-5} mbar while maintaining the substrate at a constant temperature of 650 K throughout the deposition process. Argon gas (99.99%) served as the sputtering gas, while nitrogen gas (99.99%) acted as the reactive gas. Prior to deposition, a 5 min sputtering process was executed on the zirconium target to eliminate any surface oxide. The flow rates of argon and nitrogen gases were held constant at 40 and 1 sccm, respectively, and the deposition process was conducted for 90 min.

Subsequent to the deposition of thin ZrN films with a thickness of 200 nm, they underwent diode laser irradiation at a visible wavelength of 650 nm for durations of 6 and 10 min. The laser was applied in a defocused beam mode to ensure uniform irradiation across the sample surface. The deposition parameters of the thin ZrN films used in the current investigations are given in Table 1.

Table 1. Deposition parameters of thin ZrN films.

Parameter	Chosen Value
Base pressure	2×10^{-5} mbar
Operating pressure	8.5×10^{-3} mbar
Substrate-to-target distance	60 mm
Substrate temperature	650 K
Sputtering gas (Ar:N2)	(40:1) sccm
Duty cycle	10%
Sport size	500 μm
Pulse duration	1 μs
Irradiance/laser power density	10 W/cm^2
Power pulse	19.6 W
Pulse frequency	100 kHz

These samples were characterized using a variety of techniques. The crystalline structure of the ZrN films was analyzed using X-ray diffraction (XRD) with Cu Kα radiation (λ = 1.5406 Å). XRD patterns were captured at a grazing incidence angle (3°) within the 2θ range of 30°–80°, employing a scan rate of 0.02°. The surface morphology, microstructure, and phase compositional analysis of the thin ZrN films were examined using a scanning electron microscope (Carl Zeiss AG, Oberkochen, Germany) and high-resolution transmission electron microscopy (HRTEM). The HRTEM studies were conducted out on a JEOL 2000 EXII TEM (JEOL, Tokyo, Japan) run at 200 kV. The ZrN films were initially cut into small pieces and thinned to approximately 200 nm using a focused ion beam (FIB) system. After thinning, the samples were cleaned in an ultrasonic bath to remove any surface contaminants. Finally, the thinned films were placed on a suitable support for TEM imaging and analysis. The electrical properties of the thin films were evaluated using the four-point probe technique, while their mechanical properties were measured using nanoindentation techniques. Electrical measurements were performed at room temperature under ultrahigh-vacuum conditions employing a four-tip configuration with a tip spacing of 5 μm, corresponding to a 15 μm separation between the current-carrying probes. Mechanical measurements were performed using the continuous stiffness measurement (CSM) technique on a nanoindenter. The CSM method enabled depth-controlled indentations, with a maximum indentation depth of less than 200 nm. The primary aim of the nanoindentation study was to evaluate the effects of indentation depth on the material's load, elastic modulus, and hardness. A strain rate of 0.2 s^{-1} was applied during the measurements, and Poisson's ratio of the sample was determined to be 0.27 at room temperature.

3. Results and Discussion

3.1. X-Ray Diffraction (XRD)

In Figure 1, the XRD analysis showcases a thin zirconium nitride (ZrN) film deposited on a glass substrate. XRD was used to assess the crystalline nature of the ZrN films. Initially, for as-deposited samples, a single peak of ZrN and ZrO_2 emerged at 2θ = 40.51° and 42.62°, respectively. Following 6 min of irradiation, one peak of ZrN and two peaks of ZrO_2 surfaced at 2θ = 40.51°, 42.62°, and 45.90°, respectively. Notably, with 10 min of irradiation, the presence of pronounced ZrN peaks in the diffractogram of the laser-treated surface signified the crystalline behavior of ZrN, showcasing high-intensity peaks [38].

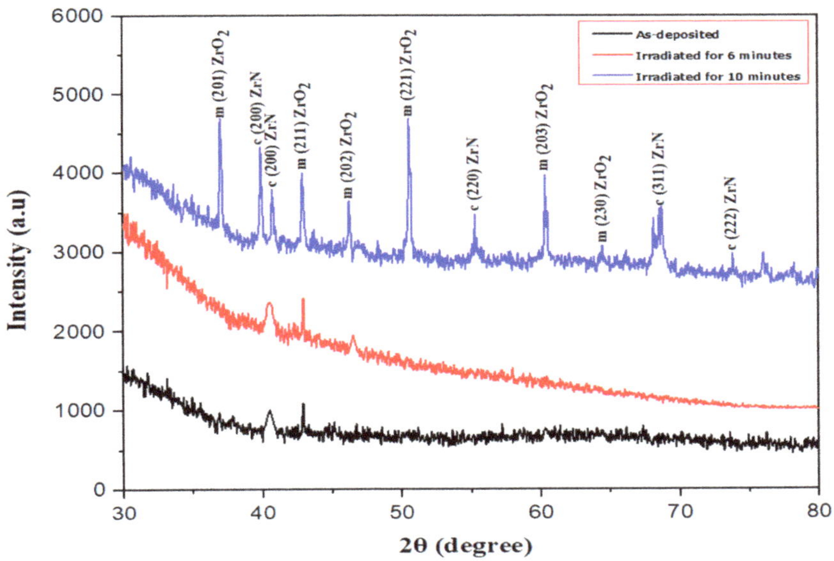

Figure 1. X-ray diffraction pattern of as-deposited/irradiated thin ZrN film. "C" and "m" indicate the cubic and monoclinic planes of ZrN and ZrO_2.

Following a 10-min treatment, a distinct reflection emerged, aligning with the ZrN (2 0 0) orientation at 39.86° and 40.51°. The pristine film's diffraction patterns reveal two peaks corresponding to stable metallic phases, specifically the cubic ZrN [cF8, Fm-3m], showcasing the (2 0 0) plane reflection at 39.86° and 40.51°. The cubic structure of the zirconium nitride film was identifiable through PDF card 02-0956. Various research groups have documented the simultaneous growth of stable ZrN phases within zirconium nitride films [28,39–42]. In the case of the sample subjected to a 10-min laser irradiation, novel peaks corresponding to cubic zirconium nitride (ZrN) with planes (2 2 0), (3 1 1), and (2 2 2) were observed at 55.26°, 68.6°, and 73°, respectively [32,43–46]. Furthermore, peaks aligning with monoclinic ZrO_2 and its planes (2 0 1), (2 1 1), (2 0 2), (2 2 1), (2 0 3), and (2 3 0) were evident at 36.22°, 42.62°, 45.90°, 50.46°, 60.33°, and 64.27°, respectively [47,48]. The monoclinic structure of the ZrO_2 film was identifiable through PDF 37-1484. The prominent ZrO_2 peaks observed in the XRD pattern after the irradiation of ZrN can be attributed to oxidation processes occurring during the laser treatment. These processes are facilitated by irradiation-induced defects, localized heating, and environmental exposure to oxygen, even under the constant argon and nitrogen flow. The fluctuation in peak intensity can be attributed to laser-induced localized heating and subsequent rapid cooling, which trigger processes such as melting and re-crystallization even in the post-deposition treatment stage. With prolonged laser irradiation, the sample progresses through phases of melting, re-solidification, and re-crystallization upon cooling. Gas atoms diffuse along grain boundaries, augmenting the concentration of gas atoms within the sample, thereby manifesting additional peaks or intensifying the peak intensity of the corresponding planes [34,38].

The calculated parameters of all fabricated materials are given in Table 2. The Bragg equation and Debye–Scherrer equations were used to calculate the interplanar spacing (d-spacing) and crystallite size.

Table 2. Summary of XRD results.

Samples	2θ (Degree)	Crystallite Size (nm)	Dislocation Density (D) m^{-2}	d-Spacing (Å)	ZrN (hkl)	Lattice Parameters (Å)
ZrN (As deposited)	40.51	31.72	0.00099	2.29	(2 0 0)	4.5674 (Å) α = 90°, β = 90°, γ = 90°
ZrN (Irradiated for 6 min)	40.51	30.46	0.001077	2.29	(2 0 0)	
ZrN (Irradiated for 10 min)	40.51	25.13	0.0015	2.29	(2 0 0)	
ZrN (Irradiated for 10 min)	55.26	175	0.000326	1.62	(2 2 0)	
ZrN (Irradiated for 10 min)	68.6	158	0.000400	1.36	(3 1 1)	
ZrN (Irradiated for 10 min)	73	133	0.000565	1.32	(2 2 2)	

The Bragg equation is as follows [49]: $n\lambda = 2dsin\theta$ or $d = \frac{n\lambda}{2sin\theta}$, where λ = 1.5406 (wavelength of incident X-rays), n = 1 (order of reflection), θ represents the peak position, and d indicates the interplanar spacing or d-spacing in (Å). The Debye–Scherrer equation is as follows [50]: $S = \frac{k\lambda}{\beta cos\theta}$, where S represents the crystallite size, k represents the Scherrer constant, and β is the measured diffraction peak at full-width half-maximum (FWHM). The number of defects in prepared samples was computed by dislocation density (D), and calculated using the equation $D = \frac{1}{S^2}$ [50–52].

Figure 2 presents a comparison of crystallite size and dislocation density of as-deposited and irradiated thin ZrN film. It can be seen in Figure 2a that the crystallite size of the prepared samples [Cubic ZrN (2 0 0)] decreased from the as-deposited state to the irradiated thin ZrN film. The average crystallite size of cubic ZrN (2 0 0) is 29.10 nm. However, the crystallite size of other ZrN peaks when irradiated for 10 min first increased and then decreased rapidly from (2 2 0) to (2 2 2) planes due to the defects [34,40,53] (Table 2). The average crystallite size of all 10 min-irradiated planes of ZrN was 92.21. The initial increase in crystallite size can be attributed to localized heating during irradiation, which promotes recrystallization and grain growth in specific regions. However, as irradiation continues, the generation of defects leads to grain refinement and a reduction in crystallite size. The laser parameters, such as power and duration, can significantly influence the evolution of crystallite size and microstructure. These findings provide valuable insights into the potential structural modifications of ZrN films induced by laser treatment. It can be seen in Figure 2b that the dislocation density of the prepared samples increased from as-deposited to irradiated thin ZrN films. The values of dislocation density show that ZrN produced more defects in the system when irradiated for 10 min [34,51,54–56]. The localized annealing and dynamic recrystallizations can increase the crystallinity, along with the dislocation density [57,58]. This can occur under specific irradiation conditions, where energy input is sufficient to reorder the crystalline structure. This dislocation density can also be confirmed with XRD graphs, as co-content has extra peaks, producing more system defects. It is clear from Table 2 that d-spacing of ZrN peaks decreased when irradiated for 10 min. Irradiation can introduce defects that initially reduce the d-spacing by altering the lattice structure. However, it can also lead to annealing effects that repair and reorganize the lattice, resulting in improved crystallinity [59,60]. Results obtained from XRD patterns were in good agreement with the results obtained from SEM and TEM analysis.

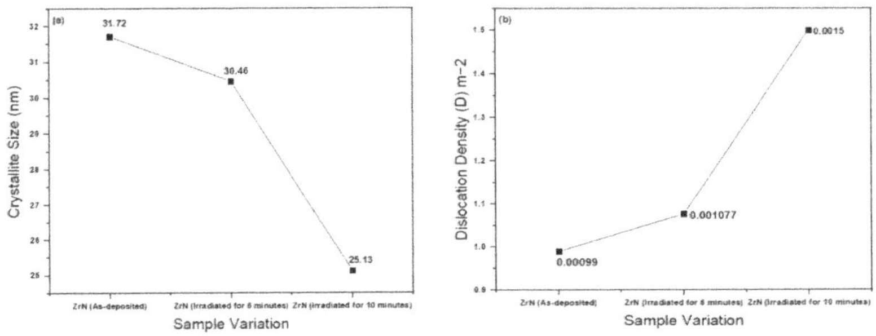

Figure 2. (**a**) Comparison of crystallite size and (**b**) dislocation density of as-deposited/irradiated thin ZrN Film.

3.2. *Scanning Electron Microscopy (SEM)*

The investigation delved into the morphology, microstructure, and chemical composition of the deposited layer to comprehend the impact of deposition conditions. Scanning electron microscopy (SEM) images were utilized to scrutinize the top surface of the treated layer, as depicted in Figure 3. The SEM image of the as-deposited ZrN sample in Figure 3a showcases high homogeneity and planarity, with some areas exhibiting small droplets on a smooth surface [61]. Figure 3b illustrates the surface after irradiation for six minutes, displaying regular ablated/melted petals due to laser scanning, leading to surface damage, cracks, and voids within the petals caused by laser irradiation. Upon irradiation for 10 min, as shown in Figure 3c, numerous voids are observed between the petals. Extended laser irradiation enhances the surface mobility of atoms within the film, leading to the formation of defects such as vacancies, interstitials, or dislocations, which in turn promote void formation. This process also contributes to increased hardness of the material. Additionally, irradiation can also induce phase transitions in the material, such as the transformation from cubic ZrN to ZrO_2, which can result in volume changes that further lead to void formation [42,62–64]. In essence, the introduction of oxygen generates voids in the thin ZrN film through irradiation-induced defect formations, oxygen diffusion, oxidation leading to ZrO_2 formation, volume expansion, and the resulting internal effects that drive void nucleation and growth. The crystallinity of the thin ZrN film was observed to increase post-irradiation.

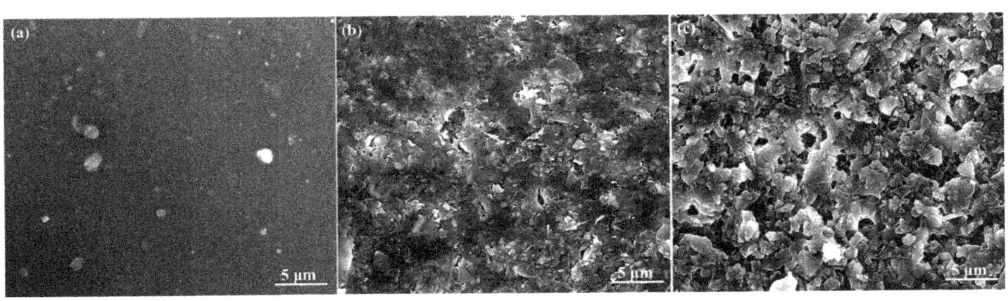

Figure 3. SEM images of as-deposited/irradiated thin ZrN film: (**a**) as deposited; (**b**) irradiated for 6 min; (**c**) irradiated for 10 min.

3.3. *High-Resolution TEM (HRTEM) Morphology*

Figure 4 presents the microstructural evolution of the as-deposited thin ZrN film and those irradiated for 6 and 10 min, denoted a1, a2, and a3, respectively. Selected area electron diffraction (SAED) analyses were conducted at the blue circle zones in b1, b2, and b3. Additionally, c1, c2, and c3 exhibit the bright-field high-resolution TEM (HRTEM) morphology of the as-deposited and 6 and 10 min-irradiated thin ZrN films. The bright-field images reveal uniformly shaped nanoscale grains and features resembling moiré fringes, a consequence of overlapping grains with varying orientations. SAED patterns indicate that the nanocrystalline grains possess random crystallographic orientations. The crystallite size of the surface layer was approximately 20 nm, 15 nm, and 10 nm for the as-deposited and irradiated thin ZrN films, respectively. Notably, the crystallite size observed via TEM was smaller than that computed using XRD, possibly due to the surface nanocrystallized layer being thinner than the detectable depth of XRD measurements [58]. Post-irradiation, the crystallite size of cubic ZrN decreased, generating defects that enhanced the crystallinity of the specimen. Decreased crystallite size correlates with increased hardness of thin films [65], as elucidated by nanoindentation techniques, and a rise in dislocation density. Smaller grains introduce more grain boundaries that impede dislocation movement, bolstering the material's strength, in line with the Hall–Petch effect [66], where reduced grain size corresponds to heightened hardness. Moreover, heightened dislocation density impedes dislocation motion, enhancing the material's resistance to deformation. The combination of reduced grain size and increased dislocation density culminated in a robust and resilient thin film.

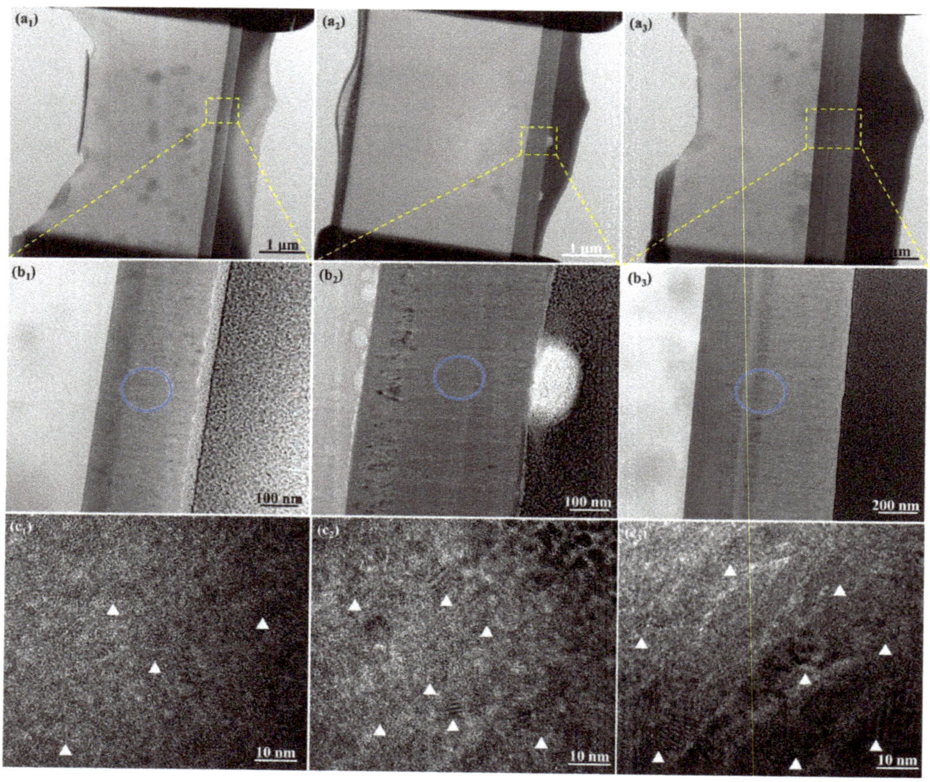

Figure 4. (**a_1–a_3**) Full TEM views of the foil; (**b_1–b_3**) magnified views of (**a_1–a_3**), blue circles indicate the zones, where selected area electron diffraction (SAED) analyses were conducted; (**c_1–c_3**) bright-field high-resolution TEM (HRTEM) images of the as-deposited 6 and 10 min-irradiated thin ZrN films, triangles indicate the regions exhibiting distinct overlapping grain features.

Figure 5 illustrates the selected area electron diffraction (SAED) patterns of the blue circled zones in Figure 4 for both the as-deposited and irradiated thin ZrN films. In these patterns,

ZrN and ZrO_2 planes are represented by green and blue, respectively. The findings reveal a close alignment between the identified planes of ZrN and ZrO_2 and the results obtained from X-ray diffraction (XRD) analyses. A laser beam's effects on a coating material depend on the material's properties, laser parameters, and environmental conditions. The apparent increase in coating thickness in the TEM images could be a result of several factors, including film reorganization or local expansion during the irradiation process. Laser irradiation can induce localized melting or softening of the film, leading to some redistribution of material, which may cause the film to appear thicker in certain areas when observed in cross section.

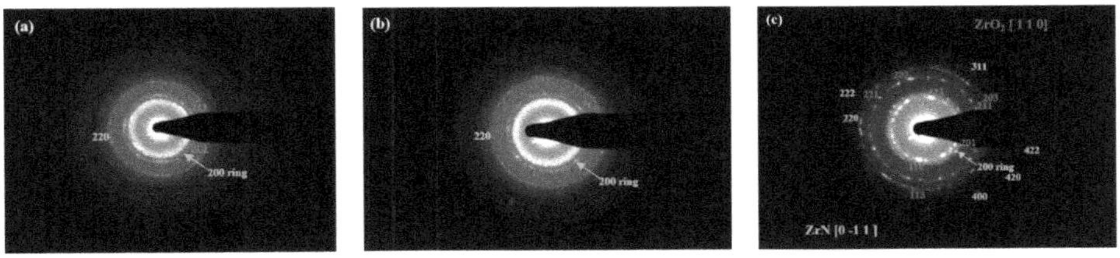

Figure 5. Selected area electron diffraction (SAED) patterns of the blue circled zone in Figure 4(b_1–b_3) for as-deposited/irradiated thin ZrN film: (**a**) as deposited; (**b**) irradiated for 6 min; (**c**) irradiated for 10 min. Blue and green show the planes of ZrN and ZrO_2.

Laser irradiation has the potential to trigger phase transformations within thin ZrN film. Specifically, under the presence of oxygen, a fraction of the ZrN material might convert to ZrO_2, either on the surface or within the film. These observations are consistent with EDX data, indicating crystalline structures that differ from the original state, signifying an enhancement in crystallinity following irradiation.

Figure 6 showcases the selected area electron diffraction (SAED) patterns of both the as-deposited and irradiated thin ZrN film. In the patterns, diffraction rings for nanosized ZrN are depicted in red, while those for nanosized ZrO_2 are shown in yellow, indicating the fabrication of nanostructured ZrN. Laser irradiation exhibited a dual role in the thin ZrN film, capable of both generating and healing defects. The high energy of the laser can introduce point defects, dislocations, or voids, while the thermal effects can simultaneously anneal existing defects, thereby reducing dislocation density. In essence, laser irradiation has the potential to significantly modify the microstructure of thin ZrN films through various mechanisms such as phase transformation, grain refinement, defect generation and annealing, alterations in surface morphology, recrystallization, oxidation, and nanoscale structural changes. These effects lead to modifications in grain size, phase composition, defect density, surface characteristics, crystallographic texture, and the presence of oxide layers [5,67].

3.4. Dispersive X-Ray Analysis (EDX)

The dispersive X-ray analysis (EDX) technique provides detailed elemental compositions of materials near the surface of a specimen, along with positional mapping [68]. Figure 7a–c display the atomic-scale morphology of the as-deposited thin ZrN film and those irradiated for 6 and 10 min, captured using aberration-corrected transmission electron microscopy (AC-TEM). The EDX spectrum revealed the presence of zirconium, nitrogen, silicon, oxygen, and carbon in the as-deposited thin ZrN film (Figure 7a). The weight percentages of Zr, N, Si, O, and C were determined to be 24.7%, 9.3%, 4.2%, 32.5%, and 29.3%, respectively. For the 6 min-irradiated ZrN (Figure 7b), the weight percentages were 25.8% Zr, 23.2% N, 1.5% Si, 37.8% O, and 11.6% C, while for the 10 min-irradiated ZrN (Figure 7c), the percentages were 29.4% Zr, 17.9% N, 1.5% Si, 39.5% O, and 11.7% C. The thin film was composed of zirconium nitride, so Zr and N were detected in the EDX spectrum as they were the primary constituents of the film. Oxygen may appear in the EDX results due to the formation of a thin oxide layer (e.g., ZrO_2) on the surface of the film. Zirconium has a strong affinity for oxygen, and when exposed to air after irradiation, it can react with oxygen to form a native oxide layer. Laser irradiation can introduce defects or induce localized heating, which can enhance the material's surface reactivity. This may increase the surface energy and allow for the absorption of oxygen from trace amounts in the environment (such as moisture or oxygen adsorbed on the surface), which would explain the increase in oxygen content observed in the EDX analysis. Copper (Cu), gallium (Ga), and platinum (Pt) also appear in the EDX results because a copper grid was used for TEM analysis. Gallium was introduced during FIB preparation, and platinum was deposited as a sputter coating to reduce sample charging and maintain stable electron interaction during TEM/EDX measurements. The EDX results across the entire zones depicted in Figure 7a–c suggest that only C and Si were dissolved in the thin ZrN films. Given that the thin film was deposited on a glass substrate, the presence of silicon in the composition is to be expected. Following sputtering, if the sample is briefly exposed to air, it can adsorb carbon-containing species from the environment. This occurrence is more likely if the sample is not promptly analyzed or stored in a controlled atmosphere, explaining the small amount of carbon detected alongside other elements. Each sample, synthesized at different irradiation times, exhibited distinct elemental compositions.

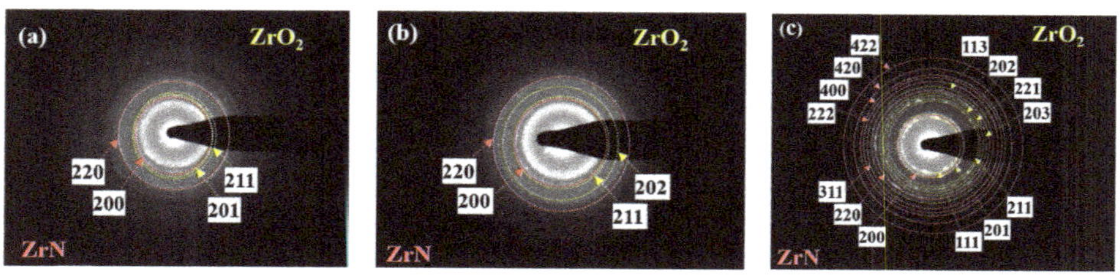

Figure 6. SAED patterns of as-deposited/irradiated thin ZrN film: (**a**) as deposited; (**b**) irradiated for 6 min; (**c**) irradiated for 10 min. Diffraction rings for nanosized ZrN (red) and diffraction rings for nanosized ZrO_2 (yellow).

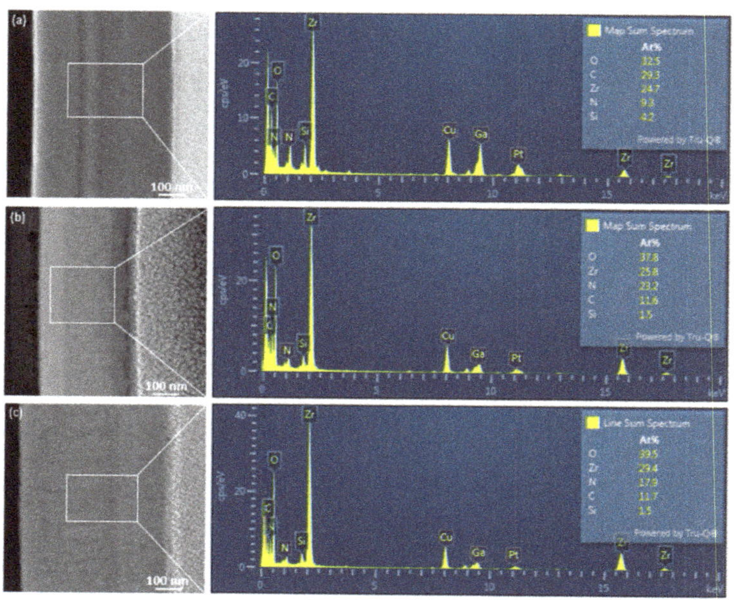

Figure 7. (**a**–**c**) Atomic-scale morphology of the as-deposited and 6 and 10 min-irradiated thin ZrN film captured by AC-TEM along with EDX spectrum.

Elemental mapping was conducted on all samples using energy-dispersive X-ray spectroscopy (TEM/EDS). The EDS results for each sample were compared to their respective targeted chemical phases, showing good agreement within the typical error margins of EDS analysis. Figure 8 displays a representative elemental mapping of as-deposited and irradiated thin ZrN film. To confirm the chemical composition, spot EDS measurements were taken on as-deposited and irradiated thin ZrN film, while area EDS was used to assess the homogeneity.

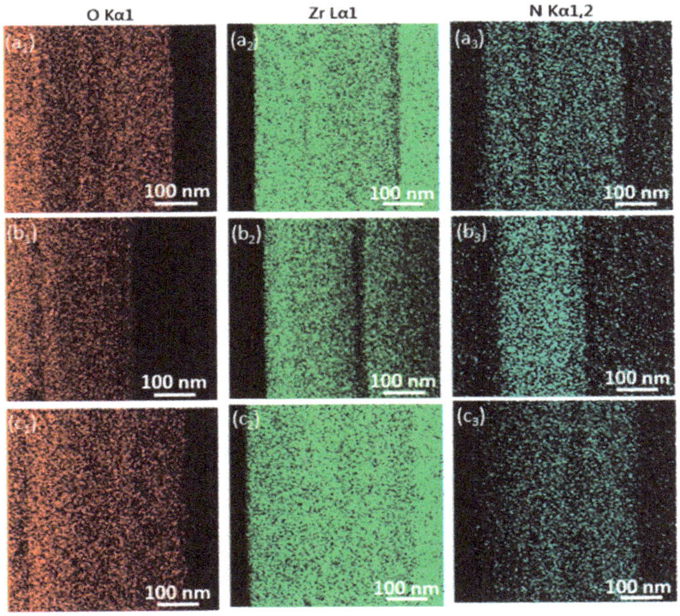

Figure 8. TEM-EDS elemental mapping images of as-deposited/irradiated thin ZrN film: ($\mathbf{a_1}$–$\mathbf{a_3}$) as-deposited; ($\mathbf{b_1}$–$\mathbf{b_3}$) irradiated for 6 min; ($\mathbf{c_1}$–$\mathbf{c_3}$) irradiated for 10 min.

Figure 9 illustrates the line-EDS profiles of zirconium (Zr), nitrogen (N), and oxygen (O) for as-deposited and irradiated thin ZrN films. These line scans corroborate the EDX atomic percentage (At. %) data, confirming the elemental distribution within the samples. In the line-EDS maps, the x-axis represents the point number along the scan path, while the y-axis indicates the intensity of each element. The thin ZrN films contained small amounts of C and Si. The presence of Si is due to the glass substrate, while carbon originates from exposure to air. Our analysis focused on ZrN and ZrO_2, so the line-EDS profiles were studied only for Zr, N, and O. The distinct profiles of Zr, N, and O reflect changes in elemental composition, influenced by the varying irradiation times applied during sample synthesis.

3.5. Nanoindentation

The nanoindentation tests were carried out utilizing the continuous stiffness measurement (CSM) indentation technique on a nanoindenter [69–71]. The CSM method features depth-controlled indentation, reaching a depth of less than 200 nm. The primary objective of the nanoindentation investigation was to analyze how depth affects the load, elastic modulus, and hardness of the material. Figures 10–12 illustrate the moduli of elasticity as a function of indentation depth, utilizing the default settings in the Agilent Nano Indenter G200 and employing the CSM method with a constant strain rate. A notable shift in the modulus of elasticity is evident for primary indentation depths below 150 nm in both the as-deposited and irradiated ZrN samples. This behavior is likely due to surface effects, such as surface oxides, which can influence the material's mechanical properties and complicate the accurate measurement or estimation of the contact area [72].

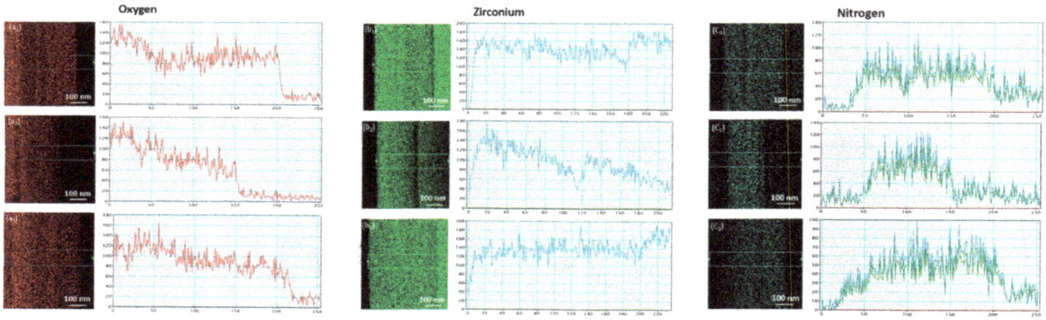

Figure 9. Line EDS of O, Zr, and N: (**a₁–c₁**) as-deposited, (**a₂–c₂**) 6 min-irradiated thin ZrN film; (a₃–c₃) 10 min-irradiated thin ZrN film.

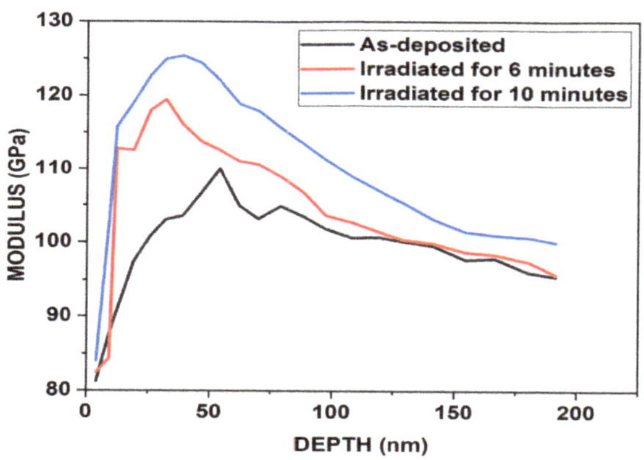

Figure 10. Elastic modulus vs. indentation depth for as-deposited and irradiated ZrN at different times.

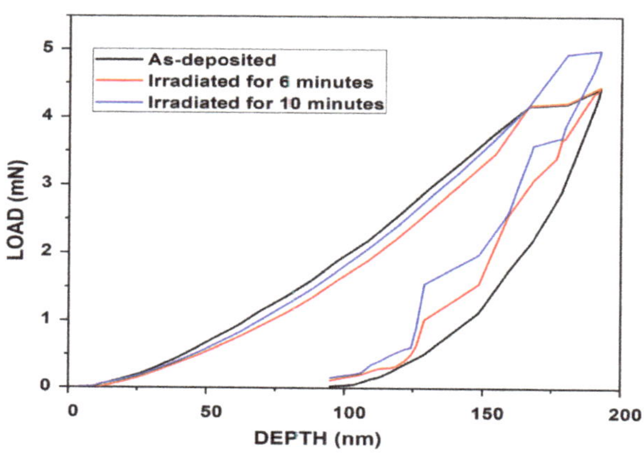

Figure 11. Load vs. indentation depth for as-deposited and irradiated ZrN at different times.

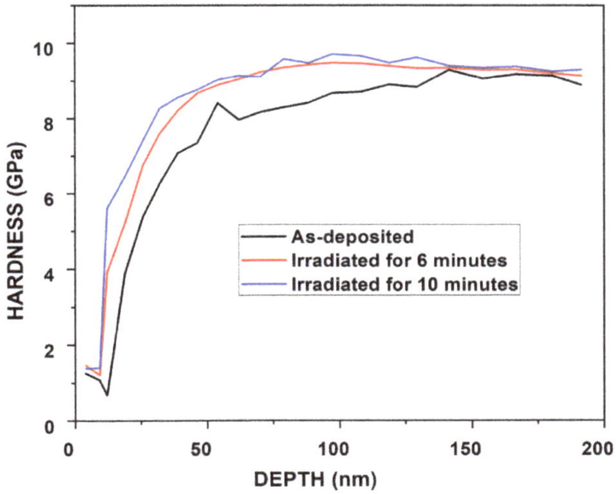

Figure 12. Hardness vs. indentation depth for as-deposited and irradiated ZrN at different times.

Nanoindentation is a widely used technique to evaluate the mechanical properties of thin films. However, it is well established that when the penetration depth exceeds 10% of the film thickness, the substrate significantly influences the measurements. This occurs because the strain field generated during indentation extends beyond the thin film into the substrate, leading to a composite response that reflects both the film and the substrate properties. In our study, while the penetration depths were carefully controlled to remain below the total thickness of the films, we acknowledge that in some cases the 10% rule may have been exceeded. This could have introduced substrate effects, potentially increasing or decreasing the measured hardness depending on the relative mechanical properties of the film and substrate. For example, a harder substrate would elevate the measured hardness, while a softer substrate would reduce it. Similarly, the elastic modulus values could represent a composite response, rather than the intrinsic properties of the thin film. To minimize such effects, we ensured consistent testing parameters, such as maintaining a low maximum load and spacing the indentations to avoid overlapping strain fields.

For the as-deposited ZrN sample, the maximum hardness was observed at a depth of 192 nm. Notably, as the hardness increased, the load and elastic modulus also peaked, as illustrated in Figures 10–12. Specifically, in the case of the as-deposited ZrN sample, the maximum hardness, load, and elastic modulus were determined to be 8.91 GPa, 4.44 mN, and 95.46 GPa, respectively. For the ZrN irradiated for 6 min, the maximum hardness, load, and elastic modulus were measured as 9.08 GPa, 4.45 mN, and 95.70 GPa, respectively. Similarly, for the ZrN irradiated for 10 min, the maximum hardness, load, and elastic modulus were found to be 9.36 GPa, 4.99 mN, and 100.03 GPa, respectively.

Following irradiation, lots of defects were created, including dislocations, interstitials, and vacancies. This increase in dislocation density hinders dislocation motion, resulting in enhanced hardness. Moreover, grain size significantly impacts hardness. Smaller grains increase the number of grain boundaries, further impeding dislocation movement and contributing to increased hardness [5,53].

The elastic modulus reflects a material's stiffness and is linked to the strength of its atomic bonds. Introducing defects such as dislocations or grain boundaries in certain cases creates "local strain fields" that enhance a material's resistance to deformation. Furthermore, structures with fine grains and higher densities of grain boundaries can enhance stiffness, as these boundaries act as reinforcement points within a material's structure [73].

In zirconium nitride (ZrN), the elastic modulus may increase due to irradiation-induced defect clusters, nanoscale grain boundaries, and dislocation networks, all of which enhance a material's mechanical properties, even in the presence of defects.

Optical microscopy was used to examine the surface and identify any visible cracks or surface features resulting from the CSM method. Figure 13a–c showcase representative optical microscopy images at the maximum indentation depth of 200 nm in the as-deposited and irradiated ZrN samples. In Figure 13a, minimal visible surface cracks or pile-ups are observed. Following irradiation for 6 min, slight alterations can be seen (Figure 13b). However, with the 10-min irradiation period, localized effects led to crack formation, as evidenced in Figure 13c [72].

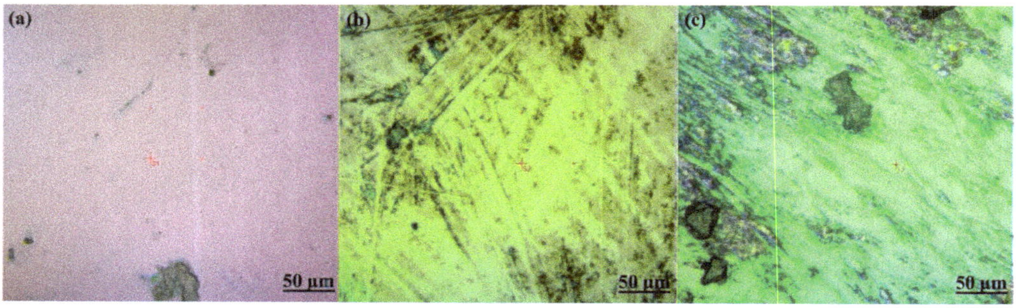

Figure 13. Optical microscopy images: (**a**) as deposited; (**b**) irradiated for 6 min; (**c**) irradiated for 10 min.

Upon comparing the results for elastic moduli, load, and hardness, a continuous increase was noted with higher laser irradiation rates, aligning well with the optical microscopy results shown in Figure 13a–c, attributed to the defects generated. Consequently, our findings suggest a hardness enhancement in the prepared thin film.

Specifically, the trend in hardness was as follows: as deposited (8.91 GPa) < irradiated for 6 min (9.08 GPa) < irradiated for 10 min (9.36 GPa).

Hence, the material's hardness increased with the rise in the laser irradiation time. This occurrence can be ascribed to the ion-exchange reaction occurring under moderate temperature and pressure conditions, acting as a rapid synthesis route for highly crystalline, stoichiometric, and nitrogen-enriched transition metal nitride alloys (TMNAs) [40,74,75].

A noteworthy observation is that the hardness and elastic moduli for the as-deposited ZrN films were lower than the typical values reported in the literature. These discrepancies can be attributed to several factors, including differences in deposition parameters such as substrate temperature, sputtering power, and pressure, which can significantly influence the microstructure of the films. The lower hardness and moduli observed in our as-deposited films may also have resulted from the deposition conditions used, which could have led to lower film quality. Furthermore, oxygen contamination and defects, as indicated by EDX analysis, might have contributed to the reduced mechanical properties of the as-deposited films.

However, following laser irradiation, significant improvements were observed in the mechanical properties of the films, with an increase in both hardness and moduli. This enhancement suggests that the laser irradiation process effectively promotes changes in the microstructure and material properties, likely through mechanisms such as defect healing, crystallization, or phase transformations, resulting in stronger and more resilient films.

3.6. Four-Point Probe

The electrical resistance of the thin film was assessed using the four-point probe method. Figure 14, illustrates the current–voltage relationships for specific thin ZrN films (as deposited and irradiated for 6 and 10 min) acquired at ambient temperature within an ultrahigh-vacuum environment. The current is plotted on the vertical axis, displaying a direct linear correlation with the applied voltage on the horizontal axis.

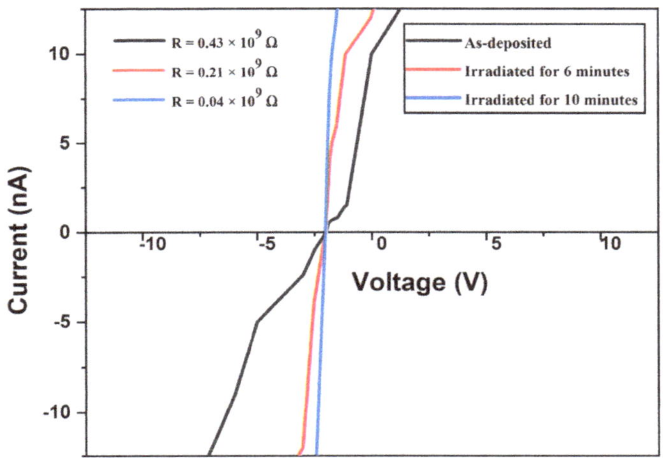

Figure 14. Electrical resistivity of as-deposited and irradiated thin ZrN films.

Upon analyzing the curves, the sheet resistances of the as-deposited ZrN and the ZrN irradiated for 6 and 10 min were determined to be 0.43×10^9 Ω, 0.21×10^9 Ω, and 0.04×10^9 Ω, respectively. After irradiation, the thin ZrN film exhibited enhanced electrical properties, characterized by a decrease in resistance [76–78].

Previous studies have revealed that the resistivity of thin ZrN films is impacted by lattice defects. As the applied bias increases, the film's resistivity gradually decreases. This decline in resistivity can be ascribed to heightened ion bombardment due to the increased bias, facilitating adatom movement towards their equilibrium positions. Consequently, the film becomes more compact, displaying reduced lattice defects. These alterations contribute to lower resistivity and enhanced crystallinity of the film.

Conversely, the occupation of interstitial positions by nitrogen atoms within the hexagonal close-packed lattice of Zr grain boundaries and lattice distortions results in improved electrical properties of the film [11,79–81]. Nevertheless, by adjusting the ZrN ratio, a transition towards the cubic ZrN phase can be induced, effectively reducing the resistivity to levels observed in cubic ZrN [82].

Our findings suggest that extending the irradiation duration decreases the electrical resistance of the thin film, thereby enhancing the electrical properties of ZrN. The availability of such high-quality samples is poised to unlock numerous possibilities in this captivating and burgeoning subfield of materials research.

4. Conclusions

In this work, it was demonstrated that laser treatment significantly influences the structural, electrical, and mechanical properties of thin ZrN films. X-ray diffraction (XRD) analysis confirmed the crystalline behavior of the thin ZrN film. Simultaneously, observations from scanning electron microscopy (SEM) analysis revealed the emergence of numerous voids within the film's surface characteristics due to laser irradiation. High-resolution transmission electron microscopy (HRTEM) examinations unveiled the formation of nanostructured ZrN with identified plane orientations, aligning well with the XRD findings. Additionally, results from the four-point probe measurements indicated that prolonged laser exposure improved the electrical properties of the thin ZrN film, as evidenced by a decrease in sheet resistance from 0.43×10^9 Ω to 0.04×10^9 Ω. Similarly, nanoindentation tests consistently portrayed an uptick in the hardness of the thin ZrN film from 8.91 GPa to

9.36 GPa, highlighting the potential of controlled laser treatment to modify and enhance coating characteristics.

Author Contributions: Conceptualization, A.N., P.L. and D.Y.; methodology, A.N. and P.L.; writing—review and editing, A.N. and P.L.; visualization, A.N.; investigation, A.N., P.L. and D.Y.; data curation, A.N.; writing—original draft preparation, A.N.; supervision, D.Y. All authors have read and agreed to the published version of the manuscript.

Funding: This research received no external funding.

Data Availability Statement: The data presented in this study can be obtained from the corresponding author on reasonable request.

Conflicts of Interest: The authors declare no conflicts of interest.

References

1. Goldschmidt, H. Nitrides. In *Interstitial Alloys*; Springer: Berlin/Heidelberg, Germany, 1967; pp. 214–253.
2. Toth, L. *Transition Metal Nitrides and Carbides*; Refractory Materials Series; Academic Press: New York, NY, USA; London, UK, 1971; p. 91.
3. Zhang, S.; Wang, J.; Wu, R.; Pan, B.; Liu, L.; Liu, C. Structural, morphological, and electrochemical capacitive properties of sputtered ZrN thin films for supercapacitor. *J. Mater. Sci. Mater. Electron.* **2023**, *34*, 1611. [CrossRef]
4. Chiang, C.; Chu, J.; Liu, F.; Liaw, P.; Buchanan, R. A 200 nm thick glass-forming metallic film for fatigue-property enhancements. *Appl. Phys. Lett.* **2006**, *88*, 131902. [CrossRef]
5. Dahmani, M.; Izerrouken, M.; Azibi, M.; Saoula, N.; Haid, F.; Sari, A.; Dahmane, A.; Ishaq, A.; Ghamnia, M. Investigation of defects formation in ZrN thin film by proton and swift heavy ion irradiations. *J. Radioanal. Nucl. Chem.* **2024**, *333*, 1097–1105. [CrossRef]
6. Liu, Y.-H.; Tai, H.-H.; Ho, C.-A.; Lin, T.-H.; Wu, M.-C.; Wang, J.-C. Highly compatible and reliable ZrN interfacial layer between TiN top electrode and antiferroelectric ZrO2 thin film to boost the electrocaloric behavior. *J. Eur. Ceram. Soc.* **2024**, *44*, 215–223. [CrossRef]
7. García, J.; Guette, A.; Medrano, A.; Labrugere, C.; Rico, M.; Lahaye, M.; Sanchez, R.; Martínez, A.; Rodríguez, R. Nitrogen ion implantation on group IVb metals:: Chemical, mechanical and tribological study. *Vacuum* **2002**, *64*, 343–351. [CrossRef]
8. Constantin, R.; Miremad, B. Performance of hard coatings, made by balanced and unbalanced magnetron sputtering, for decorative applications. *Surface Coat. Technol.* **1999**, *120*, 728–733. [CrossRef]
9. Streit, M.; Ingold, F.; Pouchon, M.; Gauckler, L.J.; Ottaviani, J.-P. Zirconium nitride as inert matrix for fast systems. *J. Nucl. Mater.* **2003**, *319*, 51–58. [CrossRef]
10. Westlinder, J.; Malmström, J.; Sjöblom, G.; Olsson, J. Low-resistivity ZrNx metal gate in MOS devices. *Solid-State Electron.* **2005**, *49*, 1410–1413. [CrossRef]
11. Larijani, M.; Tabrizi, N.; Norouzian, S.; Jafari, A.; Lahouti, S.; Hosseini, H.H.; Afshari, N. Structural and mechanical properties of ZrN films prepared by ion beam sputtering with varying N2/Ar ratio and substrate temperature. *Vacuum* **2006**, *81*, 550–555. [CrossRef]
12. Hübler, R.; Cozza, A.; Marcondes, T.; Souza, R.; Fiori, F. Wear and corrosion protection of 316-L femoral implants by deposition of thin films. *Surf. Coat. Technol.* **2001**, *142*, 1078–1083. [CrossRef]
13. Rodrıguez, R.; Garcıa, J.; Medrano, A.; Rico, M.; Sánchez, R.; Martınez, R.; Labrugère, C.; Lahaye, M.; Guette, A. Tribological behaviour of hard coatings deposited by arc-evaporation PVD. *Vacuum* **2002**, *67*, 559–566. [CrossRef]
14. Schlegel, A.; Wachter, P.; Nickl, J.; Lingg, H. Optical properties of TiN and ZrN. *J. Phys. C Solid State Phys.* **1977**, *10*, 4889. [CrossRef]
15. Sirajuddeen, M.M.S.; Banu, I.S. FP-LAPW investigation of electronic, magnetic, elastic and thermal properties of Fe-doped zirconium nitride. *AIP Adv.* **2014**, *4*, 057121. [CrossRef]
16. Cheng, D.; Wang, S.; Ye, H. First-principles calculations of the elastic properties of ZrC and ZrN. *J. Alloys Compd.* **2004**, *377*, 221–224. [CrossRef]
17. Salguero, L.; Mancera, L.; Rodríguez, J.; Takeuchi, N. First principles calculations of the ground state properties and structural phase transformation in ZrN. *Phys. Status Solidi (B)* **2006**, *243*, 1808–1812. [CrossRef]
18. Shirvani, F.; Shokri, A.; Ravan, B.A. An ab-initio study of structure and mechanical properties of rocksalt ZrN and its bilayers. *Solid State Commun.* **2021**, *328*, 114218. [CrossRef]
19. Burghartz, M.; Ledergerber, G.; Hein, H.; Van Der Laan, R.; Konings, R. Some aspects of the use of ZrN as an inert matrix for actinide fuels. *J. Nucl. Mater.* **2001**, *288*, 233–236. [CrossRef]
20. Ensinger, W.; Volz, K.; Kiuchi, M. Ion beam-assisted deposition of nitrides of the 4th group of transition metals. *Surf. Coat. Technol.* **2000**, *128*, 81–84. [CrossRef]
21. Mae, T.; Nose, M.; Zhou, M.; Nagae, T.; Shimamura, K. The effects of Si addition on the structure and mechanical properties of ZrN thin films deposited by an rf reactive sputtering method. *Surf. Coat. Technol.* **2001**, *142*, 954–958. [CrossRef]

22. Huang, J.-H.; Hsu, C.-Y.; Chen, S.-S.; Yu, G.-P. Effect of substrate bias on the structure and properties of ion-plated ZrN on Si and stainless steel substrates. *Mater. Chem. Phys.* **2003**, *77*, 14–21. [CrossRef]
23. Ribeiro, E.; Malczyk, A.; Carvalho, S.; Rebouta, L.; Fernandes, J.; Alves, E.; Miranda, A. Effects of ion bombardment on properties of dc sputtered superhard (Ti, Si, Al) N nanocomposite coatings. *Surf. Coat. Technol.* **2002**, *151*, 515–520. [CrossRef]
24. Lousa, A.; Martínez, E.; Esteve, J.; Pascual, E. Effect of ion bombardment on the properties of B4C thin films deposited by RF sputtering. *Thin Solid Film.* **1999**, *355*, 210–213. [CrossRef]
25. Thobor, A.; Rousselot, C.; Clement, C.; Takadoum, J.; Martin, N.; Sanjines, R.; Levy, F. Enhancement of mechanical properties of TiN/AlN multilayers by modifying the number and the quality of interfaces. *Surf. Coat. Technol.* **2000**, *124*, 210–221. [CrossRef]
26. Migita, T.; Kamei, R.; Tanaka, T.; Kawabata, K. Effect of dc bias on the compositional ratio of WNX thin films prepared by rf-dc coupled magnetron sputtering. *Appl. Surf. Sci.* **2001**, *169*, 362–365. [CrossRef]
27. Singh, A.; Kumar, N.; Kuppusami, P.; Prasanthi, T.; Chandramohan, P.; Dash, S.; Srinivasan, M.; Mohandas, E.; Tyagi, A. Tribological properties of sputter deposited ZrN coatings on titanium modified austenitic stainless steel. *Wear* **2012**, *280*, 22–27. [CrossRef]
28. Singh, A.; Kuppusami, P.; Khan, S.; Sudha, C.; Thirumurugesan, R.; Ramaseshan, R.; Divakar, R.; Mohandas, E.; Dash, S. Influence of nitrogen flow rate on microstructural and nanomechanical properties of Zr–N thin films prepared by pulsed DC magnetron sputtering. *Appl. Surf. Sci.* **2013**, *280*, 117–123. [CrossRef]
29. Nieh, R.E. An Evaluation of the Electrical, Material, and Reliability Characteristics and Process Viability of ZrO2 and ZrOxNy for Future Generation MOS Gate Dielectric. 2002. Available online: https://repositories.lib.utexas.edu/items/cadf3242-7eb7-4a74-adda-ae34ce9692cb (accessed on 1 December 2024).
30. Banerjee, M.; Srinivasan, N.B.; Zhu, H.; Kim, S.J.; Xu, K.; Winter, M.; Becker, H.-W.; Rogalla, D.; de los Arcos, T.; Bekermann, D. Fabrication of ZrO2 and ZrN films by metalorganic chemical vapor deposition employing new Zr precursors. *Cryst. Growth Des.* **2012**, *12*, 5079–5089. [CrossRef]
31. Tung, H.-M.; Huang, J.-H.; Tsai, D.-G.; Ai, C.-F.; Yu, G.-P. Hardness and residual stress in nanocrystalline ZrN films: Effect of bias voltage and heat treatment. *Mater. Sci. Eng. A* **2009**, *500*, 104–108. [CrossRef]
32. Yilbas, B. Laser treatment of zirconia surface for improved surface hydrophobicity. *J. Alloys Compd.* **2015**, *625*, 208–215. [CrossRef]
33. McMurtrey, M.; Sun, C.; Rupp, R.; Shiau, C.-H.; Hanbury, R.; Jerred, N.; O'Brien, R. Investigation of the irradiation effects in additively manufactured 316L steel resulting in decreased irradiation assisted stress corrosion cracking susceptibility. *J. Nucl. Mater.* **2021**, *545*, 152739. [CrossRef]
34. Egeland, G.; Valdez, J.; Maloy, S.; McClellan, K.; Sickafus, K.; Bond, G. Heavy-ion irradiation defect accumulation in ZrN characterized by TEM, GIXRD, nanoindentation, and helium desorption. *J. Nucl. Mater.* **2013**, *435*, 77–87. [CrossRef]
35. Heidarnia, A.; Ghomi, H. Helium ions irradiation of tantalum-zirconium carbide thin films: The effect of acetylene gas contents. *Radiat. Phys. Chem.* **2023**, *208*, 110927. [CrossRef]
36. Lu, F. *Phase Transformation and Microstructural Evolution of Nanostructured Oxides and Nitrides Under Ion Irradiations*; Rensselaer Polytechnic Institute: Troy, NY, USA, 2012.
37. Tarek, A.H.; Lai, C.W.; Razak, B.A.; Wong, Y.H. Physical vapour deposition of Zr-based nano films on various substrates: A review. *Curr. Nanosci.* **2022**, *18*, 347–366. [CrossRef]
38. Roman, D.; Bernardi, J.; de Amorim, C.L.; de Souza, F.S.; Spinelli, A.; Giacomelli, C.; Figueroa, C.A.; Baumvol, I.J.; Basso, R.L. Effect of deposition temperature on microstructure and corrosion resistance of ZrN thin films deposited by DC reactive magnetron sputtering. *Mater. Chem. Phys.* **2011**, *130*, 147–153. [CrossRef]
39. Khan, I.; Jabbar, S.; Hussain, T.; Hassan, M.; Ahmad, R.; Zakaullah, M.; Rawat, R. Deposition of zirconium carbonitride composite films using ion and electron beams emitted from plasma focus device. *Nucl. Instrum. Methods Phys. Res. Sect. B Beam Interact. Mater. At.* **2010**, *268*, 2228–2234. [CrossRef]
40. Mahmood, K.; Bashir, S.; Akram, M.; Hayat, A.; Rafique, M.S.; Mahmood, A. Surface, structural, electrical and mechanical modifications of pulsed laser deposited ZrN thin films by implantation of MeV carbon ions. *Nucl. Instrum. Methods Phys. Res. Sect. B Beam Interact. Mater. At.* **2019**, *448*, 61–69. [CrossRef]
41. Freitas, F.G.R.; Hübler, R.; Soares, G.; Conceição, A.G.S.; Vitória, E.R.; Carvalho, R.G.; Tentardini, E.K. Structural and mechanical properties of Zr-Si-N thin films prepared by reactive magnetron sputtering. *Mater. Res.* **2015**, *18*, 30–34. [CrossRef]
42. Rizzo, A.; Signore, M.; Mirenghi, L.; Piscopiello, E.; Tapfer, L. Physical properties evolution of sputtered zirconium oxynitride films: Effects of the growth temperature. *J. Phys. D Appl. Phys.* **2009**, *42*, 235401. [CrossRef]
43. Vasylyev, M.; Mordyuk, B.; Sidorenko, S.; Voloshko, S.; Burmak, A.; Kruhlov, I.; Zakiev, V. Characterization of ZrN coating low-temperature deposited on the preliminary Ar+ ions treated 2024 Al-alloy. *Surf. Coat. Technol.* **2019**, *361*, 413–424. [CrossRef]
44. Jelani, M.; Bashir, S.; Akram, M.; Yousaf, D.; Afzal, N.; Ahmad, S. Mechanical behaviour of excimer laser irradiated polycrystalline zirconium. *Phys. Scr.* **2014**, *89*, 025703. [CrossRef]
45. Kathuria, Y. Laser surface nitriding of yttria stabilized tetragonal zirconia. *Surf. Coat. Technol.* **2007**, *201*, 5865–5869. [CrossRef]
46. Floroian, L.; Craciun, D.; Socol, G.; Dorcioman, G.; Socol, M.; Badea, M.; Craciun, V. Titanium implants' surface functionalization by pulsed laser deposition of TiN, ZrC and ZrN hard films. *Appl. Surf. Sci.* **2017**, *417*, 175–182. [CrossRef]
47. Gorban, O.; Danilenko, I.; Nosolev, I.; Abdullayev, E.; Islamov, A.; Gavrilenko, K.; Doroshkevich, A.; Shvets, O.; Kolotilov, S. Impact of chemical and physical modification of zirconia on structure, surface state, and catalytic activity in oxidation of α-tetralol. *J. Nanopart. Res.* **2022**, *24*, 197. [CrossRef]

48. Shi, F.; Li, Y.; Wang, H.; Zhang, Q. Fabrication of well-dispersive yttrium-stabilized cubic zirconia nanoparticles via vapor phase hydrolysis. *Prog. Nat. Sci. Mater. Int.* **2012**, *22*, 15–20. [CrossRef]
49. Pope, C.G. X-ray diffraction and the Bragg equation. *J. Chem. Educ.* **1997**, *74*, 129. [CrossRef]
50. Khana, M.; Noumana, H.; Hassana, M.U.; Hasana, S.U.; Nazneena, A.; Abdelmohsenb, S.A. Structural, optical and electrical characterization of MoS2/TiO2 heterostructured thin films by chemical bath deposition technique. *Chalcogenide Lett.* **2022**, *19*, 75–82. [CrossRef]
51. Javed, U.; Sohail, H.A.; Nazneen, A.; Atif, M.; Mustafa, G.M.; Khan, M. Tuning of structural, magnetic, and optical properties of ZnO nanoparticles by Co and Cu doping. *Solid State Commun.* **2024**, *390*, 115616. [CrossRef]
52. Khan, M.; Hussain, S.; Boota, M.; Shahid, W.; Atif, M.; Nazneen, A. Optimizing the structural, optical and photovoltaic properties of Mn-doped perovskite solar cells. *Results Opt.* **2024**, *15*, 100663. [CrossRef]
53. Jiao, L.; Yu, K.; Chen, D.; Jacob, C.; Shao, L.; Zhang, X.; Wang, H. Radiation tolerant nanocrystalline ZrN films under high dose heavy-ion irradiations. *J. Appl. Phys.* **2015**, *117*, 145901. [CrossRef]
54. Calhoun, R.B.; Dunand, D.C. Dislocations in metal matrix composites: Metal matrix composites. In *Comprehensive Composite Materials: Metal Matrix Composites*; Elsevier: Amsterdam, The Netherlands, 2000; pp. 27–59.
55. Stráský, J.; Janeček, M.; Semenova, I.; Čížek, J.; Bartha, K.; Harcuba, P.; Polyakova, V.; Gatina, S. Microstructure and lattice defects in ultrafine grained biomedical $\alpha+\beta$ and metastable β Ti alloys. In *Titanium in Medical and Dental Applications*; Elsevier: Amsterdam, The Netherlands, 2018; pp. 455–475.
56. Slater, J.C. The effects of radiation on materials. *J. Appl. Phys.* **1951**, *22*, 237–256. [CrossRef]
57. Van Vuuren, A.J.; Neethling, J.H.; Skuratov, V.A.; Uglov, V.V.; Petrovich, S. The effect of He and swift heavy ions on nanocrystalline zirconium nitride. *Nucl. Instrum. Methods Phys. Res. Sect. B Beam Interact. Mater. At.* **2014**, *326*, 19–22. [CrossRef]
58. Alaneme, K.K.; Okotete, E.A. Recrystallization mechanisms and microstructure development in emerging metallic materials: A review. *J. Sci. Adv. Mater. Devices* **2019**, *4*, 19–33. [CrossRef]
59. Daghbouj, N.; Li, B.; Callisti, M.; Sen, H.; Karlik, M.; Polcar, T. Microstructural evolution of helium-irradiated 6H–SiC subjected to different irradiation conditions and annealing temperatures: A multiple characterization study. *Acta Mater.* **2019**, *181*, 160–172. [CrossRef]
60. Dolabella, S.; Borzì, A.; Dommann, A.; Neels, A. Lattice strain and defects analysis in nanostructured semiconductor materials and devices by high-resolution X-ray diffraction: Theoretical and practical aspects. *Small Methods* **2022**, *6*, 2100932. [CrossRef] [PubMed]
61. Maksakova, O.; Simoẽs, S.; Pogrebnjak, A.; Bondar, O.; Kravchenko, Y.; Beresnev, V.; Erdybaeva, N. The influence of deposition conditions and bilayer thickness on physical-mechanical properties of CA-PVD multilayer ZrN/CrN coatings. *Mater. Charact.* **2018**, *140*, 189–196. [CrossRef]
62. Dhara, S. Formation, dynamics, and characterization of nanostructures by ion beam irradiation. *Crit. Rev. Solid State Mater. Sci.* **2007**, *32*, 1–50. [CrossRef]
63. Xu, T.; Zheng, H.; Wang, J.; Banhart, F.; Sun, L. Electron Beam Irradiation Effects and In-Situ Irradiation of Nanomaterials. In *In-Situ Transmission Electron Microscopy*; Springer: Berlin/Heidelberg, Germany, 2023; pp. 17–51.
64. Ting, I.-S.; Huang, J.-H. Evaluation of incipient oxidation behavior of ZrO2-Coated Zircaloy-4 by thermogravimetric analysis. *Mater. Chem. Phys.* **2021**, *262*, 124317. [CrossRef]
65. Cavaleiro, A.; Louro, C. Nanocrystalline structure and hardness of thin films. *Vacuum* **2002**, *64*, 211–218. [CrossRef]
66. Padmanabhan, K. Mechanical properties of nanostructured materials. *Mater. Sci. Eng. A* **2001**, *304*, 200–205. [CrossRef]
67. Petrović, S.; Tsibidis, G.D.; Kovačević, A.; Božinović, N.; Peruško, D.; Mimidis, A.; Manousaki, A.; Stratakis, E. Effects of static and dynamic femtosecond laser modifications of Ti/Zr multilayer thin films. *Eur. Phys. J. D* **2021**, *75*, 304. [CrossRef]
68. Welker, R.W. Size analysis and identification of particles. In *Developments in Surface Contamination and Cleaning*; Elsevier: Amsterdam, The Netherlands, 2012; pp. 179–213.
69. Guillonneau, G.; Kermouche, G.; Bec, S.; Loubet, J.-L. Determination of mechanical properties by nanoindentation independently of indentation depth measurement. *J. Mater. Res.* **2012**, *27*, 2551–2560. [CrossRef]
70. Tabor, D. *The Hardness of Metals*; Oxford University Press: Oxford, UK, 2000.
71. Tabor, D. The hardness of solids. *Rev. Phys. Technol.* **1970**, *1*, 145. [CrossRef]
72. Jia, Y.-F.; Cui, Y.-Y.; Xuan, F.-Z.; Yang, F. Comparison between single loading–unloading indentation and continuous stiffness indentation. *RSC Adv.* **2017**, *7*, 35655–35665. [CrossRef]
73. Suryanarayana, C. Mechanical alloying and milling. *Prog. Mater. Sci.* **2001**, *46*, 1–184. [CrossRef]
74. Wang, P.; Wang, S.; Zou, Y.; Zhu, J.; He, D.; Wang, L.; Zhao, Y. Novel Nitride Materials Synthesized at High Pressure. *Crystals* **2021**, *11*, 614. [CrossRef]
75. Craciun, D.; Bourne, G.; Socol, G.; Stefan, N.; Dorcioman, G.; Lambers, E.; Craciun, V. Characteristics of ZrC/ZrN and ZrC/TiN multilayers grown by pulsed laser deposition. *Appl. Surf. Sci.* **2011**, *257*, 5332–5336. [CrossRef]
76. Shah, E.H.; Brown, B.; Cola, B.A. A study of electrical resistance in carbon nanotube–insulator–metal diode arrays for optical rectenna. *IEEE Trans. Nanotechnol.* **2017**, *16*, 230–238. [CrossRef]
77. Wojtasiak, W.; Góralczyk, M.; Gryglewski, D.; Zając, M.; Kucharski, R.; Prystawko, P.; Piotrowska, A.; Ekielski, M.; Kamińska, E.; Taube, A. AlGaN/GaN high electron mobility transistors on semi-insulating Ammono-GaN substrates with regrown ohmic contacts. *Micromachines* **2018**, *9*, 546. [CrossRef] [PubMed]

78. Zhang, L.; Austin, D.; Merkulov, V.I.; Meleshko, A.V.; Klein, K.L.; Guillorn, M.A.; Lowndes, D.H.; Simpson, M.L. Four-probe charge transport measurements on individual vertically aligned carbon nanofibers. *Appl. Phys. Lett.* **2004**, *84*, 3972–3974. [CrossRef]
79. Liu, C.-P.; Yang, H.-G. Systematic study of the evolution of texture and electrical properties of ZrNx thin films by reactive DC magnetron sputtering. *Thin Solid Film.* **2003**, *444*, 111–119. [CrossRef]
80. Ching, W.; Xu, Y.-N.; Ouyang, L. Electronic and dielectric properties of insulating Zr3N4. *Phys. Rev. B* **2002**, *66*, 235106. [CrossRef]
81. Norouzian, S.; Larijani, M.M.; Afzalzadeh, R. Effect of nitrogen flow ratio on structure and properties of zirconium nitride films on Si (100) prepared by ion beam sputtering. *Bull. Mater. Sci.* **2012**, *35*, 885–887. [CrossRef]
82. Ul-Hamid, A. The effect of deposition conditions on the properties of Zr-carbide, Zr-nitride and Zr-carbonitride coatings—A review. *Mater. Adv.* **2020**, *1*, 988–1011. [CrossRef]

nanomaterials

MDPI

Article

A Comparative Study of Neutron Shielding Performance in Al-Based Composites Reinforced with Various Boron-Containing Particles for Radiotherapy: A Monte Carlo Simulation

Shiyan Yang [1,2], Yupeng Yao [3], Hanlong Wang [1] and Hai Huang [1,*]

[1] Key Laboratory of Material Physics, Ministry of Education, School of Physics, Zhengzhou University, Zhengzhou 450001, China; shiyanyang@alu.fudan.edu.cn (S.Y.)

[2] Institute of Modern Physics, Fudan University, Shanghai 200433, China

[3] Department of Radiation Oncology, Qilu Hospital of Shangdong University, Jinan 250012, China

* Correspondence: huanghai@zzu.edu.cn

Abstract: This study aimed to assess and compare the shielding performance of boron-containing materials for neutrons generated in proton therapy and used in boron neutron capture therapy (BNCT). Five composites, including AlB_2, Al-B_4C, Al-TiB_2, Al-BN, and Al-TiB_2-BN, were selected as shielding materials, with concrete used as a benchmark. The mass fraction of boron compounds in these materials ranged from 10% to 50%. The Monte Carlo toolkit Geant4 was employed to calculate shielding parameters, including neutron ambient dose equivalent, dose values, and macroscopic cross-section. Results indicated that, compared to concrete, these boron-containing materials more effectively absorb thermal neutrons. When the boron compound exceeds 30 wt.%, these materials exhibit better shielding performance than concrete of the same thickness for neutrons generated by protons. For a given material, its shielding capability increases with boron content. Among the five materials when the material thickness and boron compound content are the same, the shielding performance for neutrons generated by protons, from best to worst, is as follows: Al-TiB_2, Al-B_4C, AlB_2, Al-TiB_2-BN, and Al-BN. For BNCT, the shielding performance from best to worst is in the following order: Al-B_4C, AlB_2, Al-TiB_2, Al-TiB_2-BN, and Al-BN. The results of this study provide references and guidelines for the selection and optimization of neutron shielding materials in proton therapy and BNCT facilities.

Keywords: boron-containing Al-based composites; neutron shielding; ambient dose equivalent; Monte Carlo simulation

1. Introduction

Recent advancements in radiotherapy technology have led to the increasing prominence and widespread adoption of emerging particle therapies, such as proton and heavy ion radiotherapy [1,2], as well as boron neutron capture therapy (BNCT) [3]. Compared to conventional X-ray radiotherapy, these techniques offer several advantages, including enhanced biological effects and improved protection of surrounding healthy tissues.

Despite these advanced treatment techniques providing many advantages, all of them face a common challenge: neutron radiation shielding [4,5]. In proton and heavy ion radiotherapy, neutrons are primarily produced through the inelastic interactions between primary particles and atomic nuclei. These neutrons, which have high energy and a broad energy distribution (e.g., for the highest energy of 250 MeV used in therapeutic protons, the energy of neutrons generated by the protons at the angle 0° relative to the beam axis ranges from a few eV to 200 MeV [4]), contribute a lower percentage to the therapeutic target dose compared to primary particles. However, their presence increases radiation dose levels for the public and occupational workers, posing a significant risk.

In BNCT, neutrons are used directly as the primary particle source for tumor irradiation. Traditionally, reactors were employed to provide neutron sources with energies around 10 MeV. Recent developments have applied accelerators for neutron generation, which have lowered both the maximum and average neutron energies, thus offering several benefits to BNCT treatment. Nevertheless, neutrons still have a maximum energy of a few MeV, continuing to present challenges for effective shielding.

Neutrons pose a significant health hazard due to their higher relative biological effectiveness and have a greater range in matter compared to charged particles of the same energy. Consequently, more effective materials are needed for adequate shielding. Currently, concrete walls are commonly used for shielding in radiotherapy settings, such as hospitals, due to their ease of processing and cost-effectiveness. However, they present several issues, including susceptibility to cracking, insufficient shielding efficiency, and the need for greater thickness to achieve adequate protection [6,7]. To meet shielding requirements, recent research has focused on boron-containing aluminum (Al)-based materials [8–10], which leverage aluminum's unique properties—such as lightweight, high strength, and high wear resistance—combined with boron's excellent neutron shielding performance. Common examples include Al-B_4C [8,9], AlB_2 [11], and Al-TiB_2 [12,13]. Titanium (Ti) is often added to enhance interfacial properties and increase the strength of aluminum [13]. These materials have been widely adopted in neutron radiation shielding, and their shielding performance for neutrons has been investigated and compared. For instance, Qi et.al. [10] reviewed the advantages and disadvantages of neutron shielding metal alloy materials, such as B_4C/Al and B/Al alloy, for the storage of spent nuclear fuel. Yusupov et.al [14] manufactured a highly borated (~50 wt.%) dispersed boron–aluminum composite and assessed its neutron-shielding properties. Jia et al. [15] investigated the neutron shielding and mechanical properties of a short carbon fiber-reinforced B_4C/Al hybrid composite. Lee et.al [16] evaluated the mechanical and thermal neutron absorbing properties of B4C/Al alloy composites.

The purpose of this study is to evaluate and compare the neutron shielding performance of boron-containing Al-based materials in radiotherapy settings, including proton therapy and BNCT, and to explore their potential applications in enhancing neutron shielding in radiotherapy. Due to the limitations of performing measurements only in specialized facilities with well-prepared shielding materials, this study employs Monte Carlo (MC) simulations to assess the shielding effectiveness.

2. Materials and Methods

All simulations presented in this work were conducted using the Geant4 (version 10.1.2) MC toolkit [17,18]. The G4EmStandardPhysics_option4 and QGSP_BIC_HP physics lists were used to handle electromagnetic and hadronic interactions, respectively. In the hadronic physic list, the neutron high precision (HP) package was applied to handle neutron interactions from thermal energies up to 20 MeV.

To assess and compare the performance of various boron-containing aluminum-based materials to neutrons in radiotherapy, two radiotherapy scenarios were selected: proton radiotherapy and BNCT. To achieve the research objectives, this study was divided into several steps, as detailed below.

2.1. Secondary Neutron Spectra Simulations

To obtain the secondary neutron spectra generated by radiotherapeutic protons, a detailed nozzle module of a proton and heavy ion accelerator, built in our previous work [19,20], was used. A water phantom ($30 \times 30 \times 40$ cm^3) was positioned at the end of the beamline, and its geometrical center aligned with the isocenter of the accelerator. Four cylindrical detectors, each with a diameter of 229 mm and a length of 330 mm, consistent with the SWENDI-2 (Thermo Scientific, Waltham, MA, USA) neutron detector [21], were applied to record the secondary neutron spectral fluence. These detectors were placed at angles of 0°, 45°, 90°, and 135° relative to the beam axis, with their centers fixed at a distance of 100 cm from the isocenter. Figure 1a shows the schematic diagram of the simulation.

It should be mentioned here that the SWENDI-2 detector is cylindrical, with a diameter of 22.9 cm and a height of 21.2 cm. It utilizes polyethylene as a moderator and a ^3He tube (2 atm) as the detector. The composites of the four simulated detectors of this study were set to air because the recorded neutron kinetics do not depend on the detector's composition. Geant4 provides built-in methods for users to obtain the current position of particles and the volume they belong to. During the simulation, once a neutron is determined to have entered the volume of the detector, its kinetic energy is recorded. This differs somewhat from experimental measurements, where the measured neutron's kinetic is inferred from the signal size produced by the neutron's interaction with the detector material. This process is directly related to the detector material, as different materials have varying cross-sections for neutron collisions, resulting in different signal sizes.

Considering that the energy distribution of secondary neutrons depends on the energy of the primary protons, three proton spread-out Bragg peaks (SOBPs) with different modulation ranges were used as a primary source. These SOBPs were optimized and implemented in our previous work [20]. Each SOBP has a fixed modulation width of 50 or 40 mm, with centers located at depths of 75 (from 50 to 100 mm, denoted as SOBP1), 175 (from 150 to 220 mm, denoted as SOBP2), and 300 mm (from 280 to 320 mm, denoted as SOBP3), respectively, covering most clinical therapy needs. The integrated depth dose (IDD) curves for these SOBPs in water are shown in Figure 1b.

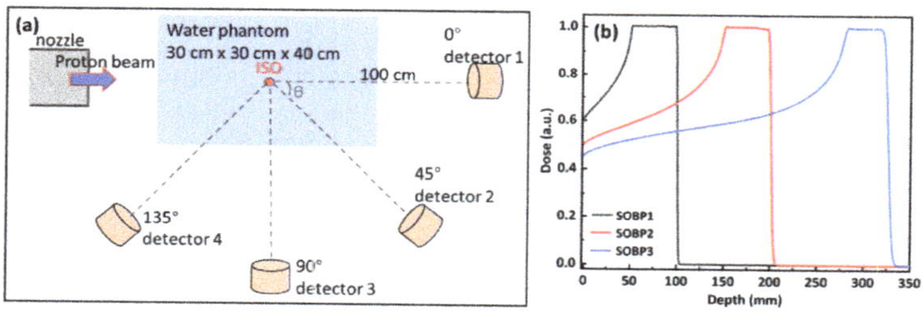

Figure 1. (**a**) Schematic diagram of the simulation setup for secondary neutrons produced by protons. (**b**) Three proton SOBPs with different modulation ranges were used as primary sources.

When a neutron entered a detector, its kinetic energy was recorded, and then it was immediately killed (stop tracking of the event in the simulation) because we were only interested in its kinetic energy upon entering the detector. A total of 1×10^8 protons were simulated for each SOBP to ensure that statistical uncertainties in the dose calculations did not exceed 1%. After the simulations were completed, the neutrons recorded by each detector were tallied in 300 logarithmic energy bins, ranging from 10^{-9} MeV to 250 MeV to obtain the secondary spectrum at a specified detector angle.

2.2. Selection of Shielding Materials

Five types of boron-containing Al-based composites were selected as neutron shielding materials: Al-B_4C, Al-BN, AlB_2, Al-TiB_2, and Al-TiB_2-BN. For convenience, each material is represented as Al-xM1-yM2, where x and y are the weight percent of additions M1 and M2, respectively. For the Al-xB_4C, Al-xBN, Al-xB_2, and Al-$xTiB_2$ compositions, x values are 10, 20, 30, 40, and 50 wt.%. For the Al-$xTiB_2$-yBN composition, x values are 10, 15, 20, and 30 wt.%, and corresponding y values are 40, 15, 30, and 20 wt.%, respectively. The weight percent of B in TiB_2 was fixed at 32%, as it has been reported that the optimal mass fraction of Ti is about (2.2 × B + 1~1.5%). [12]. The simulated weight percentages of boron cover the ranges reported in the current literature. The shielding performance of concrete was also calculated, using it as a benchmark material for comparison and analysis. Its composition and density [22] are shown in Table 1.

Table 1. Concrete composition.

Density	2.35 g cm^{-3}
Mass fraction (%)	H (1.0), C (0.1), Mg (52.91), Al (3.9), Si (33.7), Ca (4.4), K (1.3), Na (1.6), Fe (1.4)

2.3. Shielding Performance Assessment

The neutron shielding performance of different materials was evaluated in physical quantities, including ambient dose equivalent and dose recorded in the region of interest (ROI). Two types of neutron beams were used: one generated by protons and the other used in BNCT. These beams were set as primary sources and emitted vertically on the shielding material. The distance from the source to the upstream surface of the material was fixed at 100 cm, representing typical conditions for most facilities of proton and BNCT treatments. A spherical scoring volume 12 cm in diameter made of soft tissue-equivalent material (10.1 wt.% H, 11.1 wt.% C, 2.6 wt.% N, and 76.2 wt.% O, with a density of 1.0 g/cm^3) was implemented to score the neutron spectral fluence after passing through the material. This sphere is recommended by the International Commission on Radiation Units and Measurements (ICRU) for use in radiation safety and dosimetry applications [23]. For each scenario (neutron source and specified material), various material thicknesses from 50 to 350 cm were simulated. The distance between the center of the ICRU sphere and the downstream surface of the shielding material was fixed at 30 cm. To ensure uniform neutron recording, the source was modeled as a circular plane with a diameter of 15 cm, slightly larger than the scoring volume. Neutrons were uniformly sampled from the source plane and incident perpendicular to the material. The center of the source and that of the ICRU sphere were aligned with each other along the neutron incident direction (overlapped from the beam eye's view). Recorded quantities included neutron spectral fluence passing through the shielding material, neutron dose, γ dose, and total dose in the scoring volume. Neutron spectra were scored in units of cm^{-2} per incident proton. To obtain reliable statistical results for all scenarios, the number of simulated particles varied with material thickness: 1×10^7 particles for thinner materials and 1×10^8 for thicker materials.

Neutron spectral fluences recorded in the ICRU sphere were converted to the ambient dose equivalent, $H^*(10)$, using an in-house developed code and dose conversion factors provided by the International Committee of Radiological Protection (ICRP) publication 116 [24].

The ambient dose equivalent, $H^*(10)$, is calculated using the following equation [4,25]:

$$H^*(10) = \sum_{i=1}^{N} \Phi_i \, * \, h_i \tag{1}$$

where i is the ith energy bin in the neutron spectral fluence, Φ_i is the fluence in the ith energy bin, and h_i is the corresponding conversion coefficient for that energy bin.

On the other hand, the total macroscopic cross-section, Σ_t, of each shielding material was calculated using the well-known Beer–Lambert law [26]:

$$I = I_0 e^{-\Sigma_t x} \tag{2}$$

where I and I_0 are the intensity of the neutrons recorded in the ICRU sphere when the shielding material was used or not, and x is the thickness of the shielding material in cm.

3. Results

3.1. Secondary Neutron Spectral Produced by Protons

Figure 2 shows the secondary neutron spectral fluence produced by three proton SOBPs in water, as recorded by detectors placed at different locations. Each spectrum exhibits two distinct peaks: a low-energy peak, attributed to isotopic neutrons produced by evaporation processes, and a high-energy peak, associated with forward-peaked neutrons from intra-nuclear cascade reactions [4]. The ratio of the amplitude of the low-energy peak to that of the high-energy peak increases with the increasing angle between the detector and beam axis. The neutron spectra for a given SOBP are similar in shape, but as the angle increases, the amplitude of the low-energy peak increases, while that of the high-energy peak decreases, and the high-energy peak shifts towards the lower values. For different SOBPs, the energies of the high-energy peaks correlate directly with the proton energies that form the SOBP. For instance, at a 0° detector angle, the energies at the high-energy peaks are approximately 56.5 (SOBP1), 95.5 (SOBP2), and 135.5 MeV (SOBP3), respectively. Additionally, the amplitudes of both peaks increase with the primary proton energy.

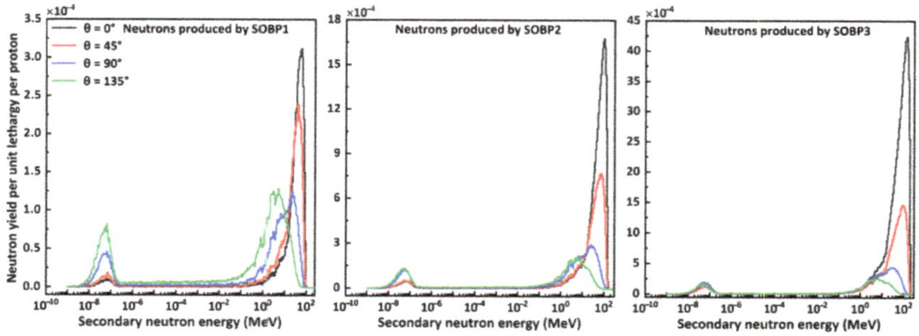

Figure 2. Secondary neutron spectral fluence produced by protons with different energies recorded at various detector angles. All figures share the same legend.

Considering the energy and intensity distribution of the secondary neutron spectra, two typical spectra were selected for evaluating the shielding performance of materials. The first spectrum, produced by SOBP1 at 135° with an average energy of 14.90 MeV, and the secondary spectrum, with the maximum energy, produced by SOBP3 at 0° with an average energy of 109.24 MeV, were referred to as n-SOBP1 and n-SOBP3, respectively, as listed in Table 1.

For BNCT scenarios, neutrons produced from different reactions serve as primary particles and have varying energies across different facilities. In this study, the neutron energy spectrum from the p-Li reaction at a proton energy of 2.8 MeV was selected as a primary source. This spectrum, shown in Figure 3, has an energy range from 0 to 1.3 MeV and is provided by Kiyanag [27].

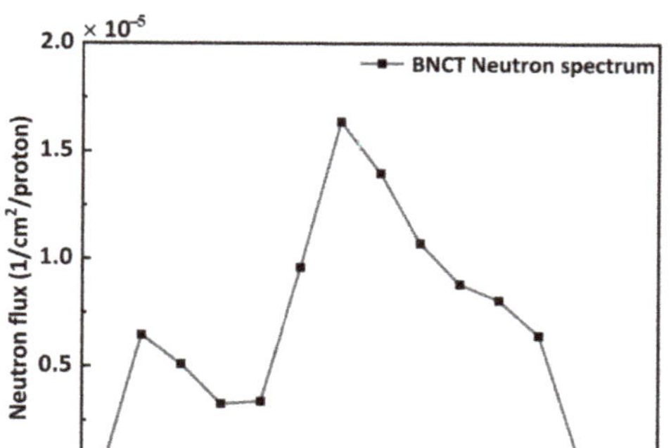

Figure 3. Neutron energy spectrum from the p-Li reaction at a proton energy of 2.8 MeV.

As a result, three neutron spectra—two produced by protons and one used in BNCT—were utilized to study the shielding performance of materials. Their details are listed in Table 2.

Table 2. Neutron spectra utilized in this work to study the shielding performance of the materials.

Label	Energy (MeV)			Description
	Min	Max	Average	
n-BNCT	0	1.30	0.65	Neutron spectrum from the p-Li reaction and unitized in BNCT
n-SOBP1	1.0×10^{-9}	67.30	14.90	Neutrons generated by SOBP1, and scored at 135°
n-SOBP3	1.0×10^{-9}	209.87	109.24	Neutrons generated by SOBP3, and scored at 0°

3.2. Neutron Spectral Fluence After Passing Through Varying Materials

Figure 4 shows neutron spectral fluence recorded in the ICRU sphere after passing through shielding materials of varying thicknesses. For convenience of comparison, the spectrals of the three studied scenarios at the source plane are displayed in the first row. The results are normalized to per primary neutron.

As demonstrated in the figure, boron-containing materials exhibit more efficient thermal neutron absorbers compared to concrete, especially for neutrons with energies below 10 keV, after passing through these boron-containing components, where almost none remain. On the other hand, the neutron fluence decreases, as expected, with increasing material thickness, either for concrete or for boron-content shielding materials. By comparing the neutron fluence distribution for a given material, it can be concluded that increasing the material's thickness offers greater improvement in shielding performance than increasing the boron content. For instance, the data reveal that for neutrons with energies ranging from 10 keV to 100 MeV, when the material thickness is increased from 50 cm to 100 cm, the fluence at a specified energy decreases by approximately two orders of magnitude for neutrons produced by SOBP1. The same increased magnitude was also observed for neutrons produced by SOBP3 when the material thickness was increased from 50 cm to 200 cm. In contrast, for those neutrons, increasing the mass fraction of boron compounds from 10% to 50% only reduces the fluence at a certain energy by at most 2 to 3 times. Specially for some composites (e.g., Al-B4C, AlB2), the increase in boron content did not result in a reduction in the fluence of high-energy neutrons (around 100 MeV of n-SOBP3).

For the neutrons used in BNCT (as shown in the first column of Figure 4), the neutron fluence recorded after passing through boron-containing components is higher than that recorded after passing through concrete with the same thickness. This phenomenon contrasts with that observed for neutrons generated by protons. One possible reason is that the neutrons used in BNCT have lower energies (below 1.5 MeV, as shown in Figure 3). These energy range neutrons primarily losing energy through elastic collisions. Low atomic number elements, such as hydrogen, have a higher elastic cross-section. Concrete contains hydrogen, which is crucial for moderating these neutrons, resulting in better shielding effectiveness.

Figure 5 shows the elastic and inelastic collision cross sections of neutrons with various elements that constitute the shielding materials, based on data from the ENDF (Evaluated Nuclear Data File, https://www.nndc.bnl.gov/endf/index.html (accessed on 23 September 2024)). The elastic collision cross section for neutrons (with energy ranges from a few eV to 2 MeV) with hydrogen is significantly larger than that for other elements. This effectively explains why concrete, which contains hydrogen, provides better shielding performance for neutrons in this energy range compared to aluminum-based boron-containing materials.

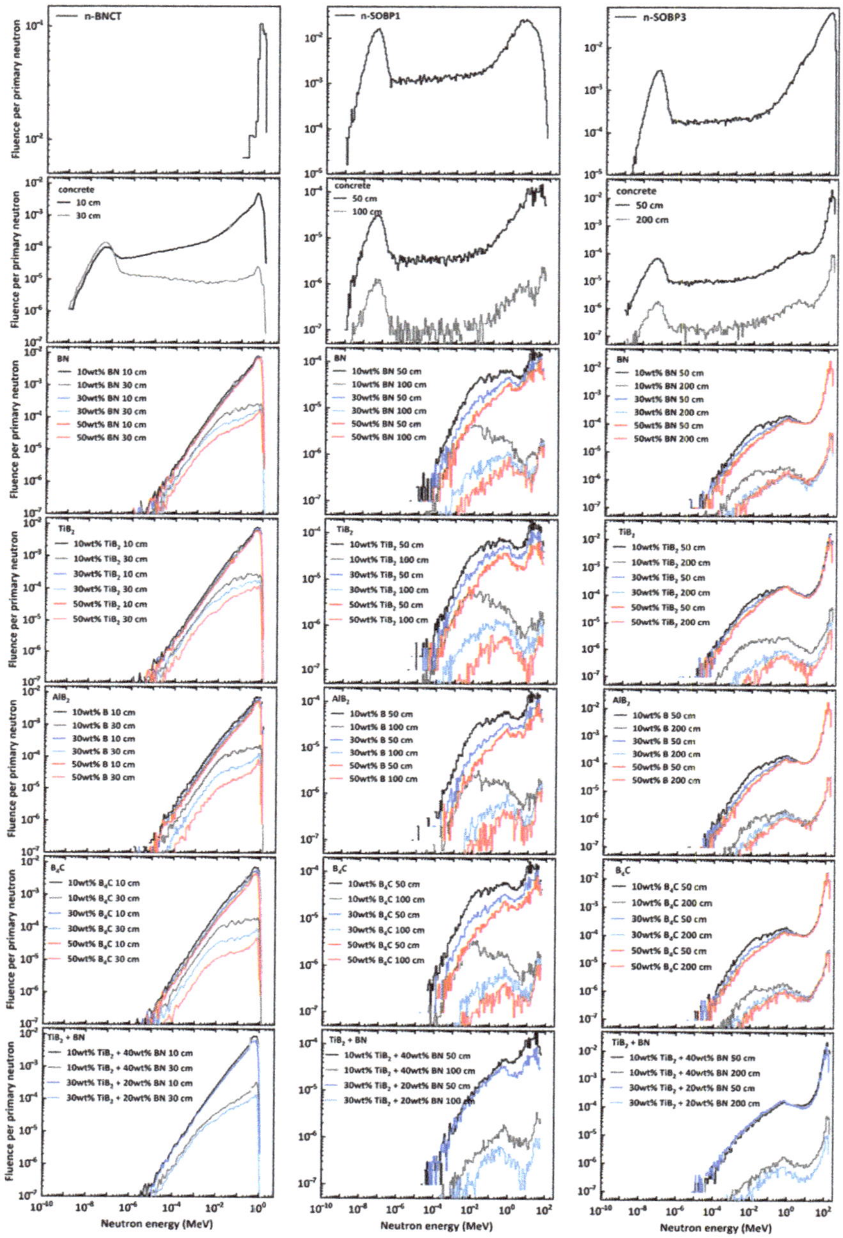

Figure 4. The neutron spectral fluence recorded in the ICRU sphere after passing through different shielding materials of various thicknesses. The first row shows the energy spectra of the three studied neutrons at the source plan.

Inelastic collisions occur only for neutrons with energies of at least 1 MeV, and the inelastic cross section of neutrons with Al is comparable to that of Si and Mg, which are the two main components of concrete.

Previous work [28] has also reported that concrete contains hydrogen, which allows it to effectively shield fast neutrons. Since neutrons and protons have almost equal masses, when a neutron collides with a hydrogen nucleus, the entire energy is transferred to the proton, resulting in the slowing down of the neutrons.

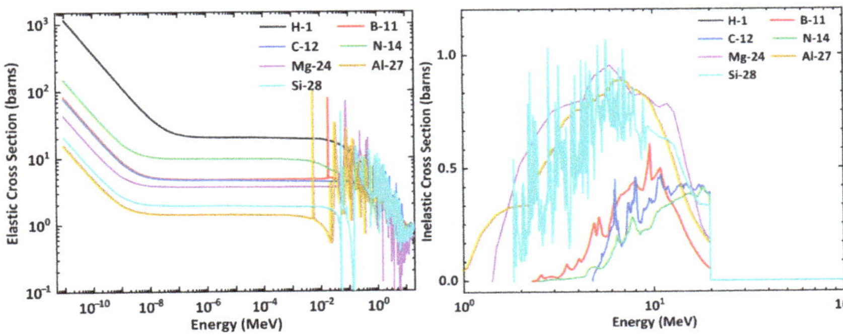

Figure 5. Elastic and inelastic collision cross sections of neutrons in different elements that constitute the shielding materials.

3.3. Neutron Ambient Dose Equivalent

For all boron-containing materials, the neutron spectral distribution and its variations with different boron content and increasing material thickness are nearly identical, as shown in Figure 4. Therefore, the neutron dose equivalent was used to quantify and compare their shielding performance.

Figure 6 shows the calculated neutron ambient dose equivalent values for different shielding materials, based on the neutron spectral fluence recorded in the ICRU sphere. All values were normalized to per incident proton. The left, middle, and right columns show the results for n-BNCT, n-SOBP1, and n-SOBP3, respectively.

For all shielding materials, the neutron ambient dose equivalent decreases sharply with increasing material thickness initially, then the reduction becomes more gradual with further increases in thickness. Once the thickness exceeds 70, 200, and 300 cm for the neutron spectra of n-BNCT, n-SOBP1, and n-SOBP3, respectively, the ambient dose equivalent value levels off, indicating that such thickness effectively shields against nearly all neutrons in these scenarios. For neutrons generated by protons (n-SOBP1 and n-SOBP3), all boron-containing materials demonstrate superior shielding performance compared to concrete with the same thickness. The neutron dose equivalent is reduced by at least approximately 50% at thicknesses of 100 and 150 cm compared to concrete for both n-SOBP1 and n-SOBP3. While for n-BNCT, the shielding performance of all boron-containing composites is inferior to that of concrete, which can be explained by the energy spectra shown in Figure 5. The neutrons have higher counts after passing through these composites than concrete because the hydrogen element in concrete moderates neutrons around the average energy of BNCT more efficiently. Moreover, for a specified material, increasing the boride compound content generally results in a lower dose equivalent value. By comparing materials with the same thickness and boron compound content, the following conclusions can be drawn: For n-SOBP1 and n-SOBP3, the neutron ambient dose equivalent decreases in the order Al-TiB_2 > Al-B_4C > AlB_2 > Al-BN. For n-BNCT, the neutron dose equivalent decreases in the order Al-B_4C > AlB_2 > Al-TiB_2 > Al/BN. For the Al-TiB_2-BN composites with the same Al mass fraction, a higher TiB_2 content results in a lower ambient dose equivalent; for example, the composite with 30 wt.% TiB_2 and 20 wt.% BN has a lower ambient dose equivalent compared to the composite with 20 wt.% TiB_2 and 30 wt.% BN.

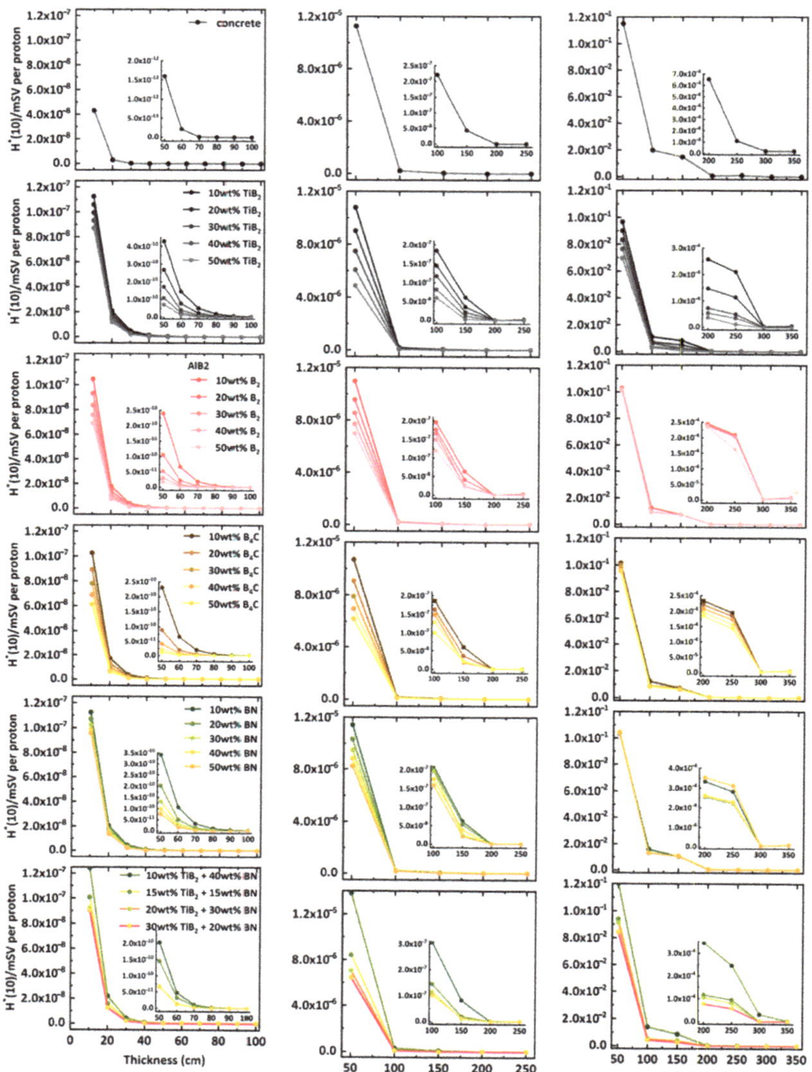

Figure 6. Neutron ambient dose equivalent values for different shielding materials. The left, middle, and right columns show the results for n-BNCT, n-SOBP1, and n-SOBP3, respectively. Each row uses the same legend.

3.4. Dose Distribution of Different Materials

Figure 7 shows the dose distribution of different particles recorded in the ICRU sphere after passing through varying thicknesses of concrete for the three studied neutron spectra. The doses were normalized per proton and categorized into three components: neutron dose, gamma dose, and others dose. The neutron and gamma doses represent the energy deposited as a result of collisions involving neutrons or photons directly participating in the interactions, respectively. While "others" dose refers to the dose contributed by secondary particles produced by neutrons (e.g., recoil nuclei from neutron inelastic collision) or by gamma rays (e.g., photoelectrons produced by the photoelectric effect), where neither neutrons nor gamma rays are directly involved in the energy deposition process. Since the dose distribution trends with increasing thickness for other materials are similar to those observed in concrete and, therefore, are not shown in the figure.

The dose distribution indicates that secondary particles contribute the most to the total dose, followed by neutrons, with gamma contributing the last. The dose contributions of neutrons and gammas are relative to the energies of the primary neutrons. For the n-BNCT, neutrons contribute approximately 25% to the total dose at a thickness of 10 cm, with their contribution gradually decreasing as thickness increases. For n-SOBP1 and n-SOBP3, neutrons contribute approximately 1% and 0.2% to the total dose, respectively, while gamma contributes approximately 3×10^{-5} and 4×10^{-6}, respectively. Moreover, for the n-SOBP1, at a thickness of 250 cm, almost all the dose is contributed by neutrons, as gamma and secondary particles are almost absorbed by the material, leaving only neutrons with high initial kinetic energy to reach the scoring sphere.

It should be noted that we categorize energy deposition into three types based on the involvement of neutrons or gammas. Although the dose from secondary particles is the largest, this does not mean we can ignore neutrons. It is important to remember that these secondary particles, including gammas, are produced by neutrons through inelastic collisions with target material; they all originate from neutrons.

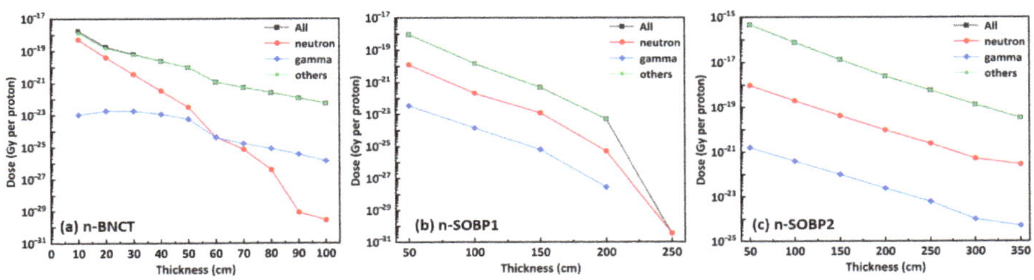

Figure 7. Dose distribution of different particles after passing through various thicknesses of concrete for the three studied neutron spectra.

Figure 8 compares the total dose distribution with increasing thickness for various materials (first column: n-BNCT, middle column: n-SOBP1, right column: n-SOBP3) scored in the ICRU sphere. For convenience of comparison, the results of concrete are plotted in red in each subplot. It should be noted that for certain thicknesses, such as 90 cm for Al-50 wt.% B_4C, total dose values are absent, as only a small number of neutrons with higher initial kinetic energy can pass through such thick material, and they are less likely to collide with the sphere to deposit energy.

The dose distribution clearly shows that, for composites of the same thickness as concrete, a lower total dose is achieved when the mass fraction of boron-containing compounds exceeds 10 wt.%, for neutrons generated by protons, and for n-BNCT when the material thickness exceeds 30 cm. The distribution also reveals that, for a given material at a specified thickness, increasing the mass fraction of a boron-containing compound from 10 wt.% to 30 wt.% results in a greater reduction in total dose compared to increasing it from 30% to 50% (especially for n-BNCT and n-SOBP1), even though both increments are by 20%. Moreover, as expected, the total dose increases with the average energy of the primary neutrons.

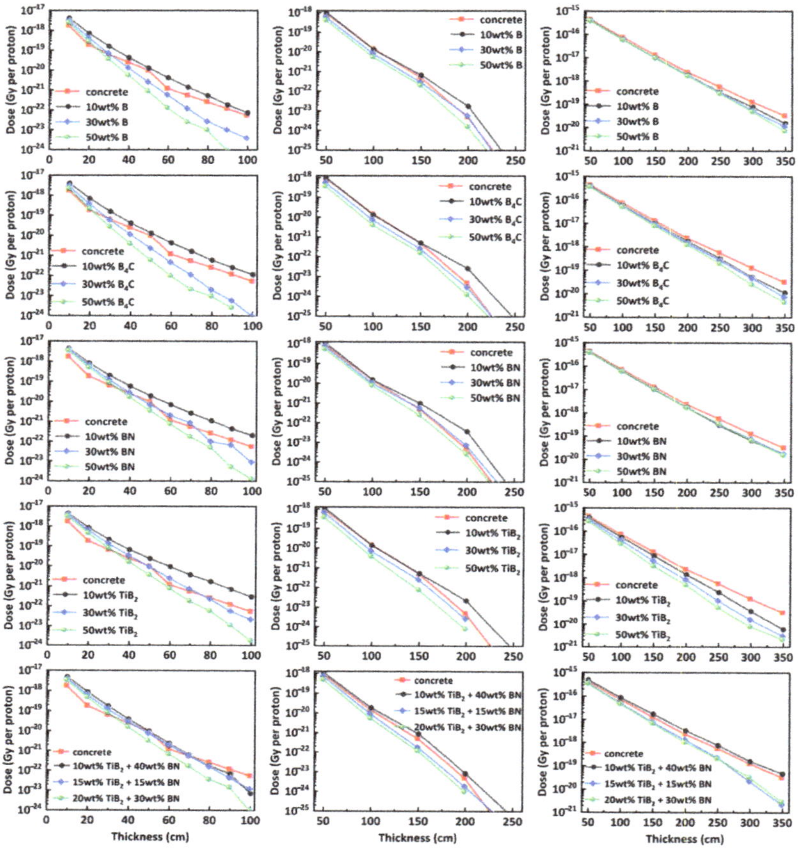

Figure 8. Comparison of total dose distribution for different materials at varying thicknesses and boron contents. The left, middle, and right columns present the results for n-BNCT, n-SOBP1, and n-SOBP3, respectively. Each row uses the same legend.

3.5. Neutron Macroscopic Cross-Sections

Figure 9 shows the neutron macroscopic cross-section for various materials with different boron contents at a fixed thickness of 10 cm for n-BNCT (top row) and 50 cm for both n-SOBP1 (middle row) and n-SOBP3 (bottom row). For comparison, the cross-section value of concrete at the same thickness is plotted as a gray line. As expected, the cross-sections increase with boron content for all three studied neutron spectra. For n-BNCT, the macroscopic cross-sections of all materials with any boron content are lower than that of concrete. In contrast, for n-SOBP1, when the boron-containing compound exceeds 30 wt.%, the macroscopic cross-sections of all materials become larger than that of concrete at the same thickness. For n-SOBP3, however, a composite with 10 wt.% boron compound content has a larger cross-section than the same concrete at the same thickness. This trend is consistent with the changes observed in the ambient dose equivalent and dose.

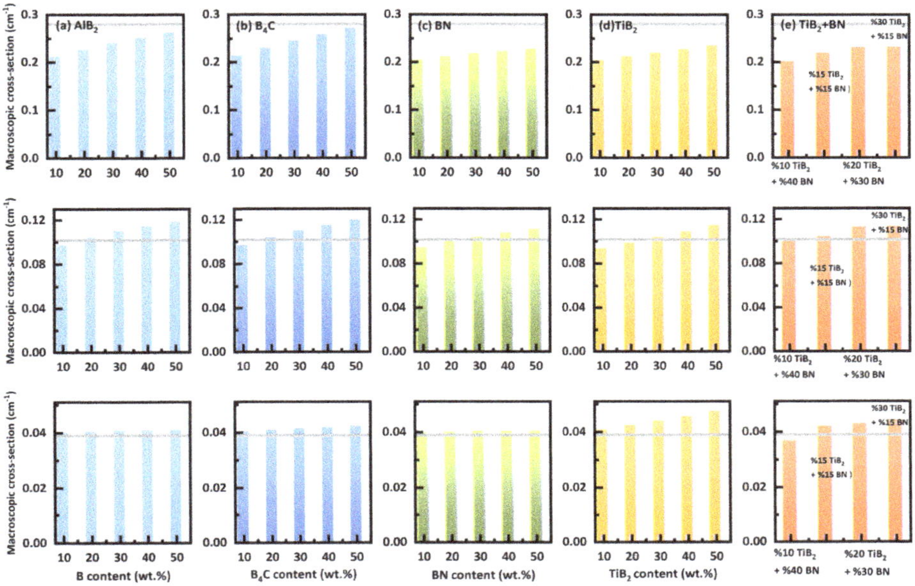

Figure 9. Neutron macroscopic cross-section for various materials with different boron contents at 10 cm for n-BNCT (**top row**) and 50 cm thickness for n-SOBP1 (**middle row**) and n-SOBP3 (**bottom row**). The gray line represents the results of concrete at the corresponding thickness. Each column uses the same legend.

4. Discussion

In this study, we evaluated the shielding performance of several boron-containing Al-based materials for neutrons in proton therapy and BNCT settings. In these treatment facilities, concrete is commonly used as shielding material. However, due to the high energy and longer range of neutrons in these environments, several meters of concrete are required to meet shielding requirements. Therefore, improving shielding efficiency and reducing the thickness of the shielding material is crucial, particularly as many additional proton and heavy ion therapy facilities are being planned and constructed. Recently, boron-containing Al-based materials (e.g., Al-B_4C, AlB_2) have been widely studied as potential replacements for concrete in neutron shielding applications. However, boron content significantly affects the microstructure of the base material. For B/Al alloy, a previous study [11] has reported that adding boron to Al tends to form boron-rich boride at the grain boundaries, which reduces the toughness of the alloy. It is well known that a high B/Al alloy containing 10 wt.% to 50 wt.% boron has an excellent neutron absorption capability; however, higher boron content also results in poor mechanical properties. For the Al-B_4C composite, the B_4C content can reach up to 50 wt.%, and it generally ranges from 5 wt.% to 50 wt.%. However, the microhardness of the composite decreases when the B_4C ratio exceeds a certain threshold [7]. Therefore, in this study, the mass fraction of boron-containing compounds was set between 10 wt.% and 50 wt.%.

Among the boron-containing Al-based materials studied in this study, Al/B_4C is the most common one, which has been widely reported in previous works and manufactured using fabrication methods [15,16]. Unlike these works, which primarily focus on evaluating the shielding performance of Al/B4C against thermal neutrons [15,16], this study investigates its shielding performance against neutrons with a broader energy distribution range. Moreover, this study employs more physical quantities, such as neutron ambient dose equivalent, cross-section, and energy spectrum, to systematically evaluate the shielding performance of these materials. By mainly applying MC simulations (theoretical calculations), this study allows for the assessment and comparison of different types of boron-containing materials simultaneously. Due to constraints such as experimental conditions, manufactures or measurements are often limited to a single material. However, manufacturing processes and mechanical properties are also crucial parameters for shielding materials, directly affecting their practical application. Therefore, in subsequent work, we will conduct experiments to study these parameters.

In our MC simulations, the shielding materials are assumed to be homogeneous, which may not fully reflect practical situations. In reality, factors such as manufacturing processes and material properties often result in shielding materials that are not perfectly homogeneous, either in density or in the distribution of added particles. However, determining the inhomogeneity of materials theoretically is challenging. Even for two samples of the same material produced under identical conditions, their uniformity can vary. Therefore, it is necessary to assume an ideal homogeneous distribution of materials when calculating their shielding performance.

In proton therapy, several types of secondary particles are generated through the nonelastic collision of protons with matter, including charged ions (e.g., helium, deuterium, tritium) and uncharged particles (e.g., gamma, neutron). These secondary ions are relatively easy to shield, requiring only a few tens of centimeters of concrete to completely absorb. As shown in Figure 1, a 230 MeV proton has a range of approximately 32 cm in water. Therefore, the primary focus of shielding in proton therapy is on uncharged particles, particularly neutrons. This study only evaluates the shielding performance of materials for neutrons, but future work should consider gamma shielding, as gamma is a non-neglect byproduct in proton therapy [29,30].

The secondary neutron spectral fluence shown in Figure 3 indicates that the shape of the spectrum depends not only on the angle relative to the beam axis but also on the energy of the primary incident proton. High-energy neutron fluences, which are the primary concern for shielding, are greatest along the beam's central axis. This suggests that if a material meets the shielding requirements at the beamline end, it will also meet the requirements in other directions. Moreover, the simulated neutron spectral fluence aligns with previous studies [4], which have demonstrated that the neutron ambient dose equivalent does not significantly change with angle relative to the beam axis in the downstream region. Therefore, when studying neutron shielding in proton therapy, it is sufficient to focus on the shielding performance along the beam's central axis.

Compared to concrete, boron-containing materials exhibit stronger absorption capabilities for thermal neutrons, as shown in Figure 4. When the boron content exceeds a certain threshold, these materials offer better shielding performance, demonstrated by lower neutron ambient dose equivalent (Figure 6) and total dose (Figure 8), as well as a higher macroscopic cross-section (Figure 9). The improved shielding performance might be attributed to the wider energy range of secondary neutrons produced by protons, from thermal to a hundred MeV. The proportion of thermal neutrons is not negligible, especially at larger angles relative to the beam axis for lower-energy incident protons. On the other hand, as neutrons pass through the shielding material, their energy decreases through collisions, continuously generating low-energy neutrons. The presence of boron in the material efficiently absorbs these thermal neutrons, resulting in dose reduction and increased cross-section.

For neutrons generated in proton therapy, this study only considers those generated in a phantom, which differs from typical patient treatment scenarios. In actual proton radiotherapy, neutrons can be generated from three processes [31–33]: (1) Production in the cyclotron, such as from proton collisions with the cyclotron walls or magnet; (2) product in the beam delivery nozzle, involving components like the range modulator wheel, ionization chambers, and window; and (3) production in the patient. Therefore, when fully simulating the proton transport process in radiotherapy, more neutrons will be produced, and more efficient shielding materials might be needed. Future work should consider neutrons produced from these various sources and processes.

5. Conclusions

This study evaluated the shielding performance of several boron-containing Al-based materials for neutrons in two scenarios: proton radiotherapy and BNCT. A total of five materials were selected, including AlB_2, Al-B_4C, Al-TiB_2, Al-BN, and Al-TiB_2-BN, each with boron compound mass fraction ranging from 10 wt.% to 50 wt.%. Neutrons generated in these two settings were set as the source. To assess and compare the shielding performance of those materials, three physical qualities—neutron ambient dose equivalent, dose contribution from different particles, and neutron macroscopic cross-section—were calculated. The results indicated that, compared to concrete of the same thickness, the absorption capacity of boron-containing materials for thermal neutrons is significantly enhanced. For neutrons generated by protons, when the mass fraction of boron compounds exceeds 30 wt.%, these materials exhibit better shielding performance than concrete, as evidenced by lower neutron ambient dose equivalents, lower dose values, as well as higher macroscopic cross-section. However, for neutrons used in BNCT, the shielding performance of all composites is inferior to that of concrete, as the hydrogen element in concrete moderates neutrons around the average energy of BNCT more efficiently. For a given material, its shielding capability increases with the boron compound content. Among the five materials when the thickness and boron compound content are the same, the shielding performance for neutrons generated by protons, from best to worst, is as follows: Al-TiB_2, Al-B_4C, AlB_2, Al-TiB_2-BN, and Al-BN. While for BNCT, the shielding performance from best to worst is in the order: Al-B_4C, AlB_2, Al-TiB_2, Al-TiB_2-BN, and Al-BN. In the next step, a more systematic computational evaluation of these shielding materials is needed to assess their performance more comprehensively. Additionally, some measurements should be conducted to validate the MC estimates.

Author Contributions: S.Y. performed the Monte Carlo simulations, helped with data analyses, interpreted the results, and drafted the paper. Y.Y. performed assistance for the Monte Carlo simulations and modifying the manuscript. H.W. assisted in modifying the manuscript. H.H. designed and supervised the project. All authors have read and agreed to the published version of the manuscript.

Funding: This work was supported by the Key Project for Science and Technology Development of Henan Province (Grant No. 242102230052), the Henan Province Postdoctoral Science Foundation (Grant No. 202102012), the Innovation Team Support Program for Cooperation of Young Talents & Enterprises in Zhengzhou University (Grant No. 32320368), and the National Training Program of Innovation and Entrepreneurship for Undergraduates (Grant No. 202410459043).

Data Availability Statement: The data that support the findings of this study are available from the corresponding author upon reasonable request.

Acknowledgments: We thank the computing resources provided by the National Supercomputing Center in Zhengzhou.

Conflicts of Interest: The authors declare no conflicts of interest.

References

1. Chen, Z.; Dominello, M.M.; Joiner, M.C.; Burmeister, J.W. Proton versus photon radiation therapy: A clinical review. *Front. Oncol.* **2023**, *13*, 1133909. [CrossRef] [PubMed]
2. Tsujii, H.; Kamada, T.; Baba, M.; Tsuji, H.; Kato, H.; Kato, S.; Yamada, S.; Yasuda, S.; Yanagi, T.; Kato, H.; et al. Clinical advantages of carbon-ion radiotherapy. *New J. Phys.* **2008**, *10*, 075009. [CrossRef]
3. Moss, R.L. Critical review, with an optimistic outlook, on Boron Neutron Capture Therapy (BNCT). *Appl. Radiat. Isot.* **2014**, *88*, 2–11. [CrossRef] [PubMed]
4. Zheng, Y.S.; Newhauser, W.; Klein, E.; Low, D. Monte Carlo simulation of the neutron spectral fluence and dose equivalent for use in shielding a proton therapy vault. *Phys. Med. Biol.* **2009**, *54*, 6943. [CrossRef]

5. Hu, G.; Hu, H.S.; Wang, S.; Han, H.T.; Otake, Y.; Pan, Z.H.; Taketani, A.; Ota, H.; Hashiguchi, T.; Yan, M.F. New shielding material development for compact accelerator-driven neutron source. *AIP Adv.* **2017**, *7*, 045213. [CrossRef]
6. Pomaro, B. A Review on Radiation Damage in Concrete for Nuclear Facilities: From Experiments to Modeling. *Mod. Simul. Eng.* **2016**, *2016*, 4165746. [CrossRef]
7. Onaizi, A.M.; Amran, M.; Tang, W.; Betoush, N.; Alhassan, M.; Rashid, R.S.; Yasin, M.F.; Bayagoob, K.H.; Onaizi, S.A. Radiation-shielding concrete: A review of materials, performance, and the impact of radiation on concrete properties. *J. Build. Eng.* **2024**, *97*, 110800. [CrossRef]
8. Gaylan, Y.; Avar, B.; Panigrahi, M.; Aygün, d.; Karabulut, A. Effect of the B_4C content on microstructure, microhardness, corrosion, and neutron shielding properties of Al–B_4C composites. *Ceram. Int.* **2023**, *49*, 5479. [CrossRef]
9. Pandiyarajan, R.; Prabakaran, M.P.; Rajkumar, T.; Vetrivel Kumar, K.; Manikandan, R. Metallurgical and mechanical properties of SiC/B_4C reinforced with aluminum composites synthesized by mechanical alloying. *Mater. Today Proc.* **2021**, *37*, 1794–1798. [CrossRef]
10. Qi, Z.D.; Yang, Z.; Li, J.P.; Guo, Y.C.; Yang, G.C.; Yu, Y.; Zhang, J.C. The Advancement of Neutron-Shielding Materials for the Transportation and Storage of Spent Nuclear Fuel. *Materials* **2022**, *15*, 3255. [CrossRef]
11. Qiao, J.; Wang, L.X.; Zhang, Q.; Yang, W.S.; Men, F.; Wu, G.H. Microstructure and mechanical properties of Bp/Al neutron shielding composites. *Nucl. Mater. Energy* **2023**, *35*, 101418. [CrossRef]
12. Grebennikov, R.V.; Chirkin, A.V.; Vukolova, V.N.; Shavrova, T.N.; Balanik, I.S. Influence of titanium on the phase composition and deformability of high boron steels. *Sov. At. Energy* **1967**, *22*, 481–486. [CrossRef]
13. Xu, G.; Wang, K.; Dong, X.; Yang, L.; Jiang, H.; Wang, Q.; Ding, W. Effects of Titanium Addition on the Microstructural and Mechanical Property Evolution of FeCrB Alloys. *Metall. Mater. Trans. A* **2020**, *51*, 4610–4622. [CrossRef]
14. Yusupov, V.S.; Tyutyunnikov, S.I.; Artyukh, V.A.; Enik, T.L.; Zelensky, V.A.; Shalyapin, V.N.; Fakhurtdinov, R.S.; Karpukhin, S.D.; Frolova, E.A.; Belelyubsky, B.F. Highly Borated Dispersed Aluminum: Experimental Evaluation of Its Neutron-Shielding Properties. *Inorg. Mater. Appl. Res.* **2022**, *13*, 823–827. [CrossRef]
15. Jia, C.P.; Zhang, P.; Xu, W.R.; Wang, W.X. Neutron shielding and mechanical properties of short carbon fiber reinforced aluminium 6061-boron carbide hybrid composite. *Ceram. Int.* **2021**, *47*, 10193–10196. [CrossRef]
16. Lee, D.; Kim, J.; Park, B.; Jo, I.; Lee, S.-K.; Kim, Y.; Lee, S.-B.; Cho, S. Mechanical and Thermal Neutron Absorbing Properties of B4C/Aluminum Alloy Composites Fabricated by Stir Casting and Hot Rolling Process. *Metals* **2021**, *11*, 413. [CrossRef]
17. Agostinelli, S.; Allison, J.; Amako, K.A.; Apostolakis, J.; Araujo, H.; Arce, P.; Asai, M.; Axen, D.; Banerjee, S.; Barrand, G.J.; et al. Geant4—A simulation toolkit. *Nucl. Instrum. Meth. A* **2003**, *506*, 250–303. [CrossRef]
18. Allison, J.; Amako, K.; Apostolakis, J.; Arce, P.; Asai, M.; Aso, T.; Bagli, E.; Bagulya, A.; Banerjee, S.; Barrand, G.J.; et al. Recent developments in Geant4. *Nucl. Instrum. Meth. A* **2016**, *835*, 186–225. [CrossRef]
19. Yang, S.Y.; Chen, B.; Zhuo, W.H.; Shen, H.; Zhao, J.F. Measurements of linear energy transfer (LET) distributions by CR-39 for a therapeutic carbon ion beam with a new 2D ripple filter. *Radiat. Phys. Chem.* **2022**, *197*, 110193. [CrossRef]
20. Yang, S.Y.; Li, Z.L.; Yao, Y.P. Evaluation of the effect of a 2D ripple filter on the averaged LET and RBE for proton and carbon ion beams: A Monte Carlo simulation study. *Eur. Phys. J. Plus* **2023**, *138*, 756. [CrossRef]
21. Ludlum Model 42–38 Wendi-2 Wide Energy Neutron Detector. 2018. Available online: https://ludlums.com/images/product_manuals/M42-38.pdf (accessed on 23 September 2024).
22. Seltzer, S.M.; Berger, M.J. Evaluation of the collision stopping power of elements and compounds for electrons and positrons. *Int. J. Appl. Radiat. Isot.* **1982**, *33*, 1189. [CrossRef]
23. ICRU. *Determination of Dose Equivalents from External Radiation Sources ICRU Report 39*; ICRU: Bethesda, MD, USA, 1985.
24. Petoussi, H.N.; Bolch, W.E.; Eckerman, K.F.; Endo, A.; Hertel, N.; Hunt, J.; Pelliccioni, M.; Schlattl, H.; Zankl, M. Conversion coefficients for radiological protection quantities for external radiation exposures. *Ann. ICRP* **2010**, *40*, 1–257. [CrossRef] [PubMed]
25. Huang, S.C.; Zhang, H.; Bai, K.; Luo, Y.; Mao, H.J.; Dai, Z.Y. Monte Carlo study of the neutron ambient dose equivalent at the heavy ion medical machine in Wuwei. *Nucl. Sci. Tech.* **2022**, *33*, 119. [CrossRef]
26. Boothroyd, A.T. *Principles of Neutron Scattering from Condensed Matter*; Oxford University Press: Oxford, UK, 2020. [CrossRef]
27. Kiyanagi, Y. Accelerator-based neutron source for boron neutron capture therapy. *Ther. Radiol. Oncol.* **2018**, *2*, 55. [CrossRef]
28. Gokul, P.; Kumar, J.A.; Preetha, R.; Chattopadhyaya, S.; Mini, K.M. Additives in concrete to enhance neutron attenuation characteristics—A critical review. *Results Eng.* **2023**, *19*, 101281. [CrossRef]
29. Verburg, J.M.; Seco, J. Proton range verification through prompt gamma-ray spectroscopy. *Phys. Med. Biol.* **2014**, *59*, 7089. [CrossRef]
30. Liu, T.; Yuan, X.; Huang, H. Primary irradiation damage in Ni–graphene nanocomposites with pre-existing hydrogen: Insights from atomistic simulations. *Eur. Phys. J. Plus* **2024**, *139*, 22. [CrossRef]
31. Prusator, M.Y.; Ahmad, S.; Chen, Y. Shielding verification and neutron dose evaluation of the Mevion S250 proton therapy unit. *J. Appl. Clin. Medical Phys.* **2018**, *19*, 305–310. [CrossRef]

32. Ma, L.; Liu, T.; Cai, B.; Liu, Z.; Zhang, P.; Li, J.; Li, H.; Huang, H. Molecular dynamics studies of primary irradiation damage in α-type Ti35 alloy. *Phys. Status. Solidi. B* **2023**, *260*, 2200560. [CrossRef]
33. Ye, Y.; Ma, L.; Tang, T.; Liu, T.; Chen, F.; Ge, X.; Jiang, Y.; Yu, X.; Huang, H. Characterization of microstructure and properties of Ti35 alloy and its high-fluence hydrogen irradiation-induced surface exfoliation. *Eur. Phys. J. Plus* **2024**, *139*, 822. [CrossRef]

nanomaterials

MDPI

Article

Microstructure Evolution of the Interface in SiC_f/TiC-Ti_3SiC_2 Composite under Sequential Xe-He-H Ion Irradiation and Annealing Process

Penghui Lei [1], Qing Chang [2,3], Mingkun Xiao [2,3], Chao Ye [2,3,*], Pan Qi [4], Fangjie Shi [5], Yuhua Hang [5], Qianwu Li [5] and Qing Peng [6,7,8,*]

1 School of Nuclear Science and Technology, Xi'an Jiaotong University, Xi'an 710049, China; penghuilei@xjtu.edu.cn
2 Institute of Clean Energy, Yangtze River Delta Research Institute, Northwestern Polytechnical University, Taicang 215400, China; changqing@mail.nwpu.edu.cn (Q.C.); xiaomingkun@mail.nwpu.edu.cn (M.X.)
3 School of Materials Science and Engineering, Northwestern Polytechnical University, Xi'an 710072, China
4 China Nuclear Power Operation Technology Corporation, Ltd., Wuhan 430223, China; qipan@cnnp.com.cn
5 Suzhou Nuclear Power Research Institute, Suzhou 215004, China; fangjie158@163.com (F.S.); hangyuhua@163.com (Y.H.); l1192177798@163.com (Q.L.)
6 State Key Laboratory of Nonlinear Mechanics, Institute of Mechanics, Chinese Academy of Sciences, Beijing 100190, China
7 Center of Materials Science and Optoelectronics Engineering, University of Chinese Academy of Sciences, Beijing 100049, China
8 Guangdong Aerospace Research Academy, Guangzhou 511458, China
* Correspondence: yechao@nwpu.edu.cn (C.Y.); pengqing@imech.ac.cn (Q.P.)

Abstract: A new type of SiC_f/TiC-Ti_3SiC_2 composite was prepared by the Spark Plasma Sintering (SPS) method in this work. The phase transformation and interface cracking of this composite under ion irradiation (single Xe, Xe + He, and Xe + He + H ions) and subsequent annealing were analyzed using transmission electron microscopy (TEM), mainly focusing on the interface regions. Xe ion irradiation resulted in the formation of high-density stacking faults in the TiC coatings and the complete amorphization of SiC fibers. The implanted H ions exacerbated interface coarsening. After annealing at 900 °C for 2 h, the interface in the Xe + He + H ion-irradiated samples was seriously damaged, resulting in the formation of large bubbles and cracks. This damage occurred because the H atoms reduced the surface free energy, thereby promoting the nucleation and growth of bubbles. Due to the absorption effect of the SiC_f/TiC interface on defects, the SiC fiber areas near the interface recovered back to the initial nano-polycrystalline structure after annealing.

Keywords: SiC_f/TiC interface; ion irradiation; phase transformation; interface cracking; bubble

1. Introduction

The accident at the Fukushima nuclear power plant revealed that metallic materials have many potential hazards under accidental working conditions, such as metal–water reactions, hydrogen decrepitation, corrosion, etc. [1,2]. In recent decades, some accident-tolerant fuel (ATF) cladding materials have been developed, such as Cr-coated zirconium alloys and FeCrAl, but the neutron economy of Cr is poor [3]. The Cr element can diffuse along grain boundaries under high-temperature steam environments, leading to cladding failure [4]. On the other hand, fiber-reinforced silicon carbide ceramic composites are also considered some of the most promising new cladding materials capable of replacing traditional alloys [5]. Compared to metallic materials, ceramic matrix composites have better hydroxide corrosion resistance and avoid hydrogen explosions, as they do not produce hydrogen gas when reacting with water under accident conditions. In addition, ceramic matrix composites are convenient to design and optimize by changing the compositions and structures of the interphase, matrix, and reinforcement materials. Boron nitride (BN) [6] and pyrolytic carbon (PyC) [7] are commonly used as interphases to enhance the toughness of SiC_f/SiC composites; however, the behaviors of these two interphase materials are unsatisfactory under irradiation conditions. Under high-temperature neutron irradiation, the volume swelling of PyC is more serious (swelling rate > 10%) [8], which can result in the formation of interface cracking [9]. On the other hand, BN is considered inadequate for nuclear applications due to its high neutron-absorbing cross-section [10]. Thus, it is necessary to find a new interfacial phase material with better irradiation tolerance.

In recent years, several novel interphases have been designed. For example, Agarwal et al. [11] reported that TiC maintains a pristine crystal structure under neutron irradiation at doses of up to 2 dpa at temperatures of 220 °C, 620 °C, and 1150 °C. Therefore, the use of TiC as an interfacial phase material may enhance the irradiation resistance of SiC_f composites. Inspired by the properties of TiC, titanium carbide coating was successfully in situ synthesized on silicon carbide fibers to serve as a novel interfacial phase in our previous work [12]. In addition, due to their unique layered structure and excellent irradiation tolerance, a novel class of ternary layered nitrides or carbides, known as MAX phase materials, has also been considered as a potential material for nuclear reactors [13–15]. As a typical MAX phase, Ti_3SiC_2 has good fracture toughness and good chemical compatibility with silicon carbide [16]. Thus, Ti_3SiC_2 could also be designed as a new interfacial material for SiC_f composites. Many prior studies [17,18] have reported various methods for preparing SiC_f composites with Ti_3SiC_2 as an interphase or matrix material. Unfortunately, the irradiation effects of these novel interface structures have not been studied in detail.

In this study, SiC_f/TiC-Ti_3SiC_2 composites were prepared by SPS to evaluate the irradiation resistance of the SiC_f/TiC-Ti_3SiC_2 interface structure. SPS, as a field-assisted sintering technology (FAST), has been applied for fabricating nuclear materials [19–25]. Ion irradiation, an economical and efficient method, was used to introduce irradiation damage into the SiC_f/TiC-Ti_3SiC_2 interface. The microstructure evolutions of the SiC_f/TiC-Ti_3SiC_2 interface structure under three different irradiation conditions (Xe, Xe + He, Xe + He + H) and the subsequent 900 °C annealing process were analyzed and have been discussed at length.

2. Materials and Methods

2.1. Materials

The reinforcement used in this study was the 3rd polymer-derived SiC fibers with 0.45 wt% oxygen content. Silicon powder (99.8 wt%) and titanium powder (99.8 wt%) were purchased from Shanghai Aladdin Biochemical Technology Co., Ltd., Shanghai, China; sodium chloride, graphite, and potassium chloride were purchased from Shanghai Macklin Biochemical Technology Co., Ltd., Shanghai, China. The materials are summarized in Table 1. Ti_3SiC_2 powders were synthesized using the molten salt synthesis method, and SiC_f/TiC-Ti_3SiC_2 composites were prepared using the SPS method. Compared to conventional unpressurized sintering reactions, molten salts can accelerate atom diffusion to enhance the ceramic sintering reaction, which could accelerate the nucleation of products and reduce the reaction temperature. Simultaneously, large amounts of molten salts can form a liquid environment to control the composition of the products [17,26–28].

Table 1. Chemicals used in this study.

Materials	Purity	Supplier
Ti	99.8%	Aladdin lnc., Shanghai, China
Si	99.9%	Aladdin lnc., Shanghai, China
graphite	99.95%	Macklin Inc., Shanghai, China
NaCl	99.8%	Macklin Inc., Shanghai, China
KCl	99.9%	Macklin Inc., Shanghai, China
ethanol	99.70%	Sinopharm Chemical Reagent Co., Ltd., Shanghai, China

2.2. Preparation of SiC_f/TiC-Ti_3SiC_2 Interface

Ti, Si, and C powders were first pre-mixed in a molar ratio of 3:1:2. Subsequently, each mole of the Ti-Si-C pre-mixed powder was combined with NaCl-KCl in a 1:1 molar ratio. The final powder mixtures were mixed in a planetary ball mill with zirconia grinding balls and ethanol for 24 h. After drying and sieving, the powder was placed in an alumina crucible, heated to 1300 °C in a high-vacuum furnace at a ramp rate of 5 °C/min, and then held at 1300 °C for 3 h. Ti_3SiC_2 powders were obtained after washing and drying. Following this, a 10% volume fraction of short-cut silicon carbide fibers and the obtained Ti_3SiC_2 powders were mixed and placed into a graphite die with an inner diameter of 10 mm; then, the sintering was conducted using an SPS facility at 1300 °C under a pressure of 40 MPa for 10 min. SiC_f/TiC-Ti_3SiC_2 composites were obtained after cooling the sinter chamber to room temperature.

2.3. Ion Irradiation

Firstly, 5 MeV Xe^{23+} ions were pre-implanted into the samples at room temperature with a fluence of 3.5×10^{15} ions/cm^2 using a 320 kV ion implanter at the Institute of Modern Physics, Chinese Academy of Sciences. Subsequently, the 400 keV He^+ and 200 keV H^+ ion irradiation studies, with fluences of 1×10^{17} ions/cm^2 and 6×10^{16} ions/cm^2, respectively, were conducted at room temperature using an NEC 400 kV ion implanter at Xiamen University. The energies of the ion beams were carefully selected to ensure that the peak concentrations and displacement damage of Xe, He, and H were located in adjacent regions. The ion incidence direction was perpendicular to the sample's surface. The displacement threshold energies for the elements Ti, Si, and C were set to 25 eV, 15 eV, and 28 eV, respectively, in the Stopping and Range of Ions in Matter (SRIM) 2013 software, based on full damage cascades mode [29]. The density of SiC and TiC was set to 3.23 g/cm^3 and 4.92 g/cm^3 in the SRIM calculations, respectively. According to Figure 1, which presents the comprehensive simulation results for the concentrations of He, Xe, and H atoms in SiC and TiC and the depth of damage, the peak hydrogen concentration, peak helium concentration, and peak displacement damage are all located at a depth of around 1 μm. Several irradiation samples were protected with argon gas (purity: 99.999%) and annealed at 900 °C for two hours in a tube furnace. Table 2 outlines the specific experimental conditions employed in this study.

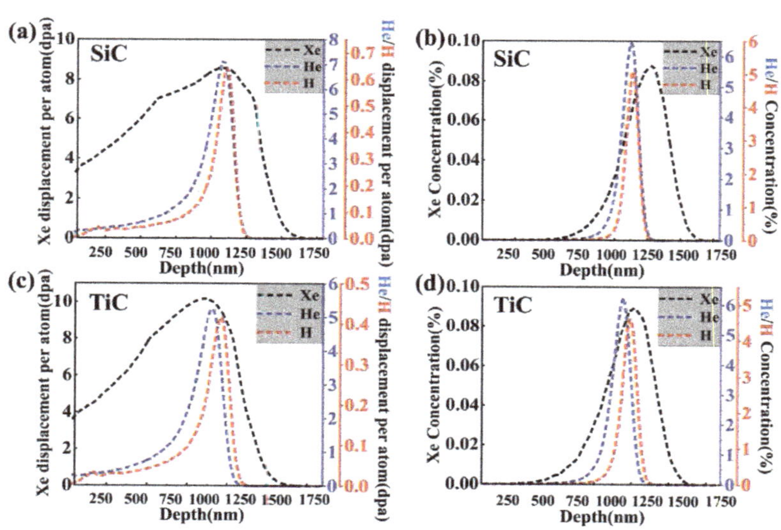

Figure 1. The results of dpa and concentrations (H, Xe, and He) in SiC and TiC calculated by SRIM 2013.

Table 2. The detailed ion irradiation experimental conditions used in this work.

Influence (Ions/cm^2)			Temperature for Annealing (°C)	Time for Annealing (h)
Xe Ions	He Ions	H Ions		
3.5×10^{15}			900	2
3.5×10^{15}	1×10^{17}			
3.5×10^{15}	1×10^{17}	6×10^{16}		

2.4. Characterization Methods

Scanning transmission electron microscopy (STEM) with the Talos F200X and scanning electron microscopy (SEM) with the Zeiss Gemini 460 were used to characterize the surface morphologies and microstructures of these unirradiated samples. The phase constitution was analyzed by X-ray diffraction (XRD) using the Bruker D8 ADVANCE at a scan rate of 10°/min. The cross-sectional TEM samples, with a thickness of less than 100 nm, were prepared using a focused ion beam (FIB) with the FEI Helios G4, employing the lift-out technique. In the initial lift-outs, 30 kV Ga^+ ions with a 9.3 nA current were used. To minimize surface damage to the samples throughout the thinning process, the current of the Ga^+ ion beam and the energy were gradually reduced, with 2 kV Ga^+ ions used for the final thinning. The damaged regions are indicated in the overview image of the TEM sample prepared by FIB.

3. Results

3.1. Characterization of SiC_f/TiC-Ti_3SiC_2 Composites

The XRD pattern of the as-prepared Ti_3SiC_2 powders is observed in Figure 2a which indicates that the primary phase in the powders is Ti_3SiC_2, with small amounts of TiC impurities. The phase compositions of the as-received SiC_f/TiC/Ti_3SiC_2 composites synthesized by SPS are shown in Figure 2b. The peaks at 10°, 20°, 39.6°, and 40.8° correspond to the (002), (004), (104), and (008) crystal planes of Ti_3SiC_2. Some peaks at 35.9°, 41.7°, and 60.5° correspond to the (111), (020), and (022) crystal planes of TiC, respectively. Diffraction peaks of the SiC fibers are located at 30° and 60°, overlapping with the diffraction peaks of TiC. Small amounts of $TiSi_2$ were also detected. After sintering, the diffraction peaks of TiC were sharp and intense, which indicates that a highly crystalline TiC phase was generated during the sintering process. The polished surface of the as-received composites is shown in Figure 2c, revealing that the SiC fibers are uniformly distributed throughout the Ti_3SiC_2 matrix, which exhibits good densification. In addition, the Ti_3SiC_2 matrix and the SiC fibers are well bonded. The inserted images in Figure 2c show the corresponding EDS mapping results for this region; the matrix and fiber regions can be clearly distinguished via the distribution of the Ti and Si elements. The irradiated and unirradiated areas, as well as the fiber and matrix regions, can also be easily distinguished by the differences in contrast in Figure 2d.

Figure 3a presents a high-angle annular dark field (HAADF) image of the interface region along with its EDX mapping results. A poor-Si zone with a thickness of about 150 nm can be seen close to the SiC fiber. The diffraction rings in Figure 3b indicate that the fibers consist of nano-SiC grains. The SAED pattern and the EDX spectrum in Figure 3b illustrate that the composition of this poor-Si zone is the TiC phase with an FCC structure. The HRTEM image in Figure 3b displays a terraced interface structure, with the TiC region near the interface containing numerous stacking faults and nano-twins.

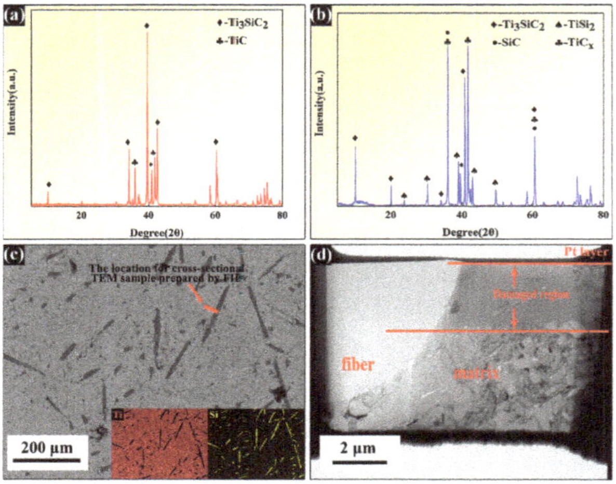

Figure 2. The XRD patterns of the synthesized Ti_3SiC_2 powders (**a**) and the SiC_f/TiC-Ti_3SiC_2 composites (**b**); (**c**) the SEM image of the surface morphologies of the as-received composites and the corresponding EDS mappings; (**d**) the overview TEM image of the TEM sample prepared by FIB.

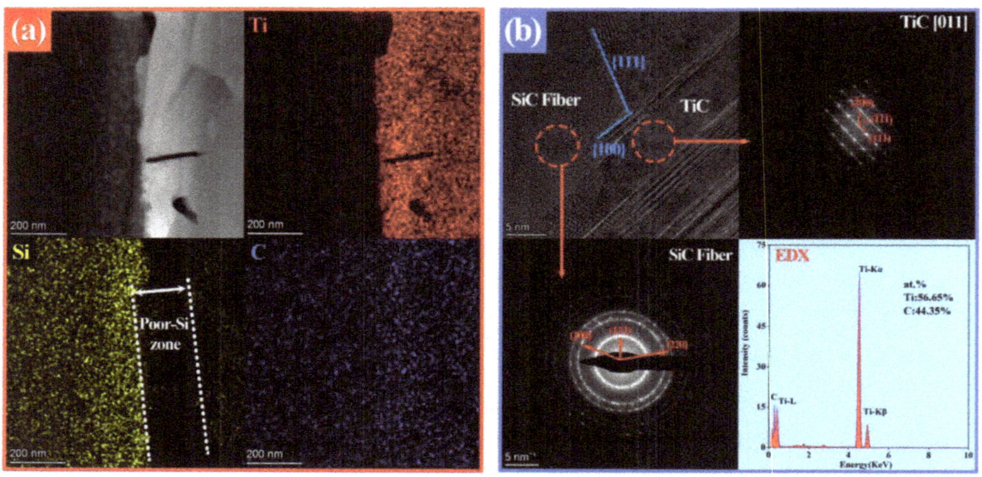

Figure 3. (**a**) The HAADF image and EDX mappings of the SiC_f/TiC interface; (**b**) the HRTEM image and FFT patterns of the SiC_f/TiC interface, with the composition near the SiC fibers shown in the EDX spectrum.

3.2. Microstructure Evolutions of the Irradiated SiC_f/TiC Interface before and after Annealing

The HRTEM images of the SiC_f/TiC interface areas with peak damage under three different irradiation conditions (Xe + He + H, Xe + He, and Xe) are shown in Figure 4a–c. The SAED patterns of the SiC fibers in these three ion-irradiated samples reveal that the SiC fibers are completely amorphized. Under the irradiation of Xe + He and Xe ions, the terraced interface morphologies are still clearly distinguishable. The density of nano-twinning structures and stacking faults in TiC is significantly higher than that in the unirradiated samples. Conversely, under the Xe + He + H irradiation condition, the terraced structure of the SiC_f/TiC interface is hardly recognizable, which indicates that the implanted H ions could destroy the terraced structure. Following a two-hour annealing at 900 °C, the microstructure of the SiC_f/TiC interface regions under three irradiation conditions is shown in Figure 5a–c. In the samples irradiated with single Xe ions, the interface remains intact, with no cracks observed, indicating that the injection of single Xe ions has minimal impact on the integrity of the interface after the annealing process. However, small bubbles and cracks were observed in the interface area of the samples injected with H and He ions. Particularly, with the introduction of H ions, a series of cavities were formed at the interface, which indicates that the implanted H ions resulted in more serious damage during the annealing process. Figure 5d–f show the SAED patterns of areas A1, A2, and A3 marked in Figure 5a–c, which indicate that SiC fibers near the interface have recrystallized. Meanwhile, the central areas of the SiC fibers, which are away from the interfaces in these three irradiated composites, are still in an amorphous state after annealing, as shown in Figure 5g–i.

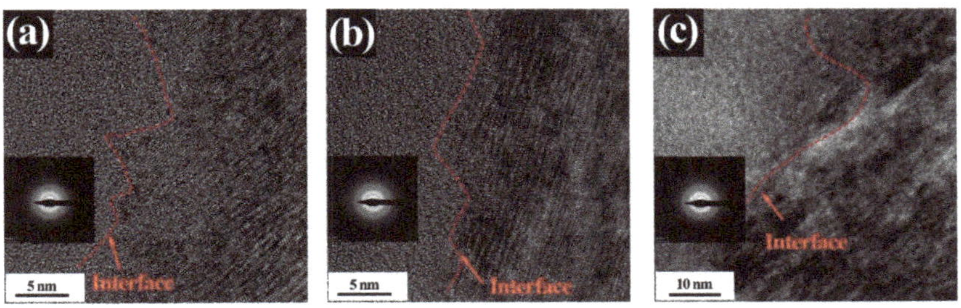

Figure 4. HRTEM images of the SiC_f/TiC interface regions. Xe irradiation (**a**); Xe + He irradiation (**b**); Xe + He + H irradiation (**c**). Each image includes the SAED patterns of the SiC fiber.

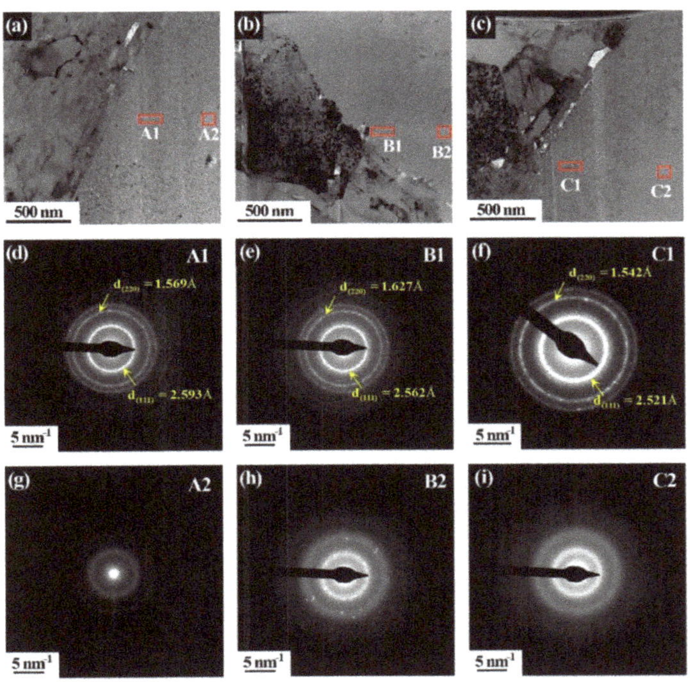

Figure 5. TEM images of irradiated samples post-annealing process: (**a**) irradiation with Xe; (**b**) irradiation with Xe + He; (**c**) irradiation with Xe + He + H; (**d–i**) the SAED patterns of A1, B1, C1, A2, B2, and C2 areas marked in (**a–c**), respectively.

4. Discussion

4.1. The Decomposition of Ti_3SiC_2 and the Formation of TiC Coatings

The decomposition of Ti_3SiC_2 can be explained by two important factors. Firstly, the high-temperature and high-pressure environment rendered Ti_3SiC_2 unstable, leading to the decomposition of Ti_3SiC_2. Compared to Ti–C bonds, the Si–Ti bonds are easier to break. As a result, Si atoms have a tendency to diffuse from the pristine lattice, causing the Ti_3SiC_2 phases to decompose into TiC during the sintering process [30,31]. Secondly, graphite molds are capable of reacting with Ti_3SiC_2 at high temperatures and pressures, and the infiltration of carbon can intensify the decomposition process, as shown in the following reaction: $Ti_3SiC_2 + (3X - 2)C \rightarrow TiC_X + Si(g)$ [32]. The SiC_f/TiC interface region was formed by several reactions: the high-temperature/-pressure environment and SiC fibers promoted the decomposition of Ti_3SiC_2 into TiC(s), Si(g), and Ti(g); then, Ti reacted with the SiC fibers to form TiC films. Meanwhile, partial Si atoms diffused into the matrix and reacted with free Ti atoms to form $TiSi_2$. These processes can be expressed as the following reactions: $Ti_3SiC_2\ (s) \rightarrow TiC_X(s) + Ti(g) + Si(g)$; $SiC_f(s) + Ti(g) \rightarrow TiC_X(s) + Si(g)$ [33].

4.2. The Microstructure Evolution of Irradiated SiC_f/TiC Interface Structure

In an irradiation environment, the irradiation damage could interplay with the interfacial interactions to control the grain boundary motions [34], which is also the reason for the structure evolution of the heterogeneous SiC_f/TiC interfaces. Firstly, during the thermal phase of the cascade evolution process, the incident ions caused atomic rearrangement and localized melting as rapid ion–atom collisions heated and melted the local lattice. Some atoms may have been displaced outside the thermal areas of the cascades and turned into interstitial defects following core solidification, even though most atoms in the cascade's core were resolidified into the crystal lattice. Then, those interstitial defects accumulated at the boundaries and changed the morphology of localized regions and the roughness of grain boundaries. The roughness became more pronounced after multiple ion impact events. Due to the extremely high migration rate of hydrogen atoms within the lattices, the implanted hydrogen ions, in addition to introducing a small amount of irradiation damage, also accelerated the aggregation of interstitial atomic defects in the interface region, which in turn exacerbated the interface coarsening.

4.3. The Phase Transformation and Cracking of Irradiated SiC_f/TiC Interface Structure after 900 °C Annealing

Zinkle et al. [35] illustrated that temperature plays a crucial role in the microstructural evolution of irradiated samples. At high temperatures, irradiation-induced vacancies and point defects are compounded and annihilated, resulting in a large reduction in the number of defects within the sample and a return of the amorphous structure to its original crystalline structure. The phase interface can act as a strong absorption sink for irradiation-induced defects. Thus, during the annealing process, numerous defects induced by irradiation, such as vacancy clusters, hydrogen atoms, and helium atoms, are prone to accumulate in the interface region. As a result, the SiC fiber regions near the interface returned to their original nano-polycrystalline structure, while other regions of the SiC fiber remained amorphous after annealing since the defects did not have enough time to migrate to the interface region. Hobbs et al. [36] reported that the amorphization of ceramics with covalent bonding is a common response to high-density defect perturbation induced by irradiation and is strongly related to the structural topology. At last, when the total number of defects reached a certain threshold, cracking formed at the interface region.

On the other hand, as Xe atoms have a large atomic radius and a high migration energy barrier, most of them were pinned in the SiC/TiC grains and grain boundaries. Therefore, no cavities or Xe bubbles were observed at the interface region in the single Xe ion-irradiated sample. With subsequent high-dose helium ion irradiation, helium bubbles prefer to nucleate at the interface region to form large bubbles [37]. Compared with the bubbles in the Xe + He ion-irradiated SiC fibers, there is a significant increase in the bubble density after the implantation of H ions, as illustrated in Figure 5b,c. The diffusion coefficient dependence on temperature is generally expressed as follows [38]: $D = D_0 \exp\left(-\frac{Q_A}{kT}\right)$, where D_0 is the constant; Q_A represents the activation energy for diffusion, which is positively correlated with atomic size; and k and T are the Boltzmann constant and temperature, respectively. Therefore, the increased bubble density following the implantation of H ions is related to the higher diffusion coefficient of H compared to those of He and Xe. Our previous study [39] illustrates that the implanted H ions have a fast migration rate inside the lattice and can attach to He bubbles to accelerate the migration of He bubbles to the interface; in addition, helium bubbles nucleate and grow more readily because H ions can reduce their surface energy [40]. The nucleation of bubbles is affected by the combination of volume free energy and surface energy, which can be described by the following equation [41]: $\Delta G = \frac{4}{3}\pi r^3 \Delta G_v + 4\pi r^2 \gamma$, where ΔG is the total free energy change; ΔG_v represents the free energy per unit volume, which becomes negative during nucleation; r represents the radius of an embryo, and when the embryo radius is larger than the critical nucleus radius r^*, nucleation is triggered; and γ represents the specific surface energy, serving as a retardation force with a surface energy of $4\pi r^2 \gamma$. Due to the reduction in surface energy induced by implanted H ions, the critical nucleus radius r^* of the bubbles decreased in the sample irradiated with Xe + He + H ions, which means that the nucleation and growth of helium bubbles become easier. Thus, the average size of bubbles in the interface region under Xe + He + H ion irradiation is larger than that in samples irradiated with only Xe + He or Xe ions. Figure 6 presents a schematic of bubble aggregation and migration in the interface region of the sample irradiated with Xe + He + H ions. Hydrogen bubbles are formed by the reaction between incident H ions and electrons from the sample [42,43], as represented by the following chemical reaction formula: $H_2 = 2H^+ + 2e^-$.

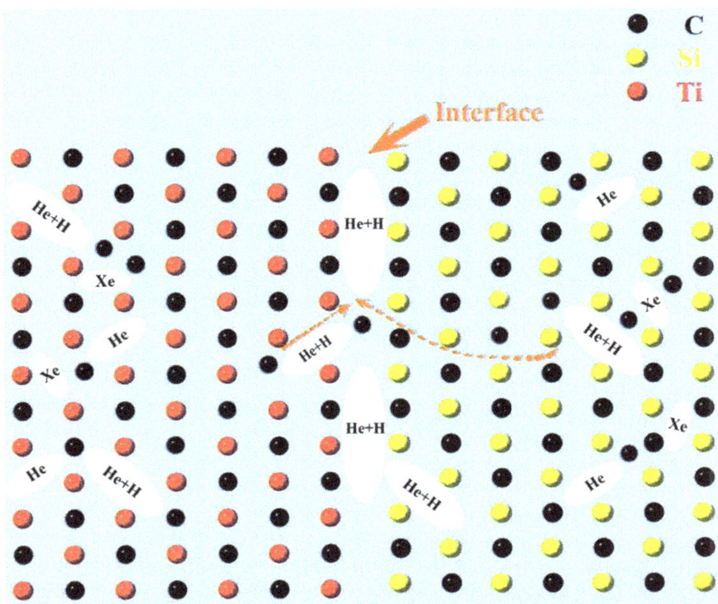

Figure 6. Schematic of bubble aggregation and migration in the interface region of the SiC_f/TiC sample irradiated with Xe + He + H ions after annealing.

The results of this work indicate that He and H atoms have a more serious impact on the material interface, and the annealing process is helpful for the recrystallization recovery of the irradiation-induced amorphous interface regions. Overall, this work suggests that a material's resistance to radiation can be enhanced through rational interface design. Our future work will focus on improving the radiation swelling and cracking resistance of the interface in SiC fiber composite materials.

5. Conclusions

SiC_f/TiC-Ti_3SiC_2 composites prepared using the SPS method were irradiated with single Xe, Xe + He, and Xe + He + H ions. Then, the microstructure evolutions of the SiC_f/TiC interface before and after annealing were investigated. The following summarizes the primary conclusions:

(1) The behaviors of the SiC_f/TiC interface under three different ion irradiation conditions (single Xe, Xe + He, and Xe + He + H) were researched. After irradiation, the SiC fibers became completely amorphous, while partial amorphization and high-density stacking faults were observed in TiC. The implanted H ions introduced more damage, thereby exacerbating the interface coarsening.

(2) After 900 °C annealing, the implanted He and H ions preferred to accumulate in the SiC_f/TiC interface region and formed bubbles. The H ions reduced surface free energy, thereby promoting the nucleation and growth of bubbles, which caused cracking. Due to the strong absorption effect of the interface on irradiation defects, the SiC fiber regions near the interface returned to their original nano-polycrystalline structure, which suggests that the materials' resistance to radiation can be enhanced through rational interface designs.

Author Contributions: Conceptualization, P.L. and C.Y.; methodology, Q.C. and Q.L.; software, M.X.; formal analysis, P.Q.; investigation, F.S. and Y.H.; writing—original draft preparation, P.L.; writing—review and editing, C.Y. and Q.P.; funding acquisition, C.Y. All authors have read and agreed to the published version of the manuscript.

Funding: This work is supported by the National Key R&D Program of China (Grant No. 2022YFB-3707200), the National Natural Science Foundation of China (No. 12205236), the Strategic Priority Research Program of the Chinese Academy of Sciences (Grant No. XDB0620103), the National Natural Science Foundation of China (Grant No. 12272378), and the High-level Innovation Research Institute Program of Guangdong Province (Grant No. 2020B0909010003).

Data Availability Statement: The data presented in this study are available on request from the corresponding authors.

Conflicts of Interest: Author Pan Qi was employed by the company China Nuclear Power Operation Technology Corporation, Ltd. The remaining authors declare that the research was conducted in the absence of any commercial or financial relationships that could be construed as a potential conflict of interest.

References

1. Ni, N.; Hudson, D.; Wei, J.; Wang, P.; Lozano-Perez, S.; Smith, G.; Sykes, J.; Yardley, S.; Moore, K.; Lyon, S. How the crystallography and nanoscale chemistry of the metal/oxide interface develops during the aqueous oxidation of zirconium cladding alloys. *Acta Mater.* **2012**, *60*, 7132–7149. [CrossRef]
2. Mazères, B.; Desgranges, C.; Toffolon-Masclet, C.; Monceau, D. Experimental study and numerical simulation of high temperature (1100–1250 °C) oxidation of prior-oxidized zirconium alloy. *Corros. Sci.* **2016**, *103*, 10–19. [CrossRef]
3. Kam, D.H.; Lee, J.H.; Lee, T.; Jeong, Y.H. Critical heat flux for SiC-and Cr-coated plates under atmospheric condition. *Ann. Nucl. Energy* **2015**, *76*, 335–342. [CrossRef]
4. Yamamoto, K.; Takayama, T.; Minamino, Y.; Koizumi, Y.; Tokunaga, T.; Hagihara, K. Modification of grain boundary microstructure by controlling dissolution behavior of θ particles in Cr-containing hypereutectoid steel. *Mater. Charact.* **2023**, *205*, 113241. [CrossRef]
5. Zinkle, S.J.; Terrani, K.A.; Gehin, J.C.; Ott, L.J.; Snead, L.L. Accident tolerant fuels for LWRs: A perspective. *J. Nucl. Mater.* **2014**, *448*, 374–379. [CrossRef]
6. Shan, Q.; Xu, Q.; Xue, Y.; Lian, J.; Wang, Y.; Chen, C.; Zhou, Y.; Ma, Q.; Shui, A. The tensile damage behavior of SiC_f/SiC–B4C after oxidation in wet atmosphere based on acoustic emission pattern recognition. *J. Am. Ceram. Soc.* **2021**, *104*, 4131–4144. [CrossRef]

7. Fellah, C.; Braun, J.; Sauder, C.; Sirotti, F.; Berger, M.-H. Influence of the carbon interface on the mechanical behavior of SiC/SiC composites. *Compos. Part A Appl. Sci. Manuf.* **2020**, *133*, 105867. [CrossRef]
8. Snead, L.; Burchell, T.; Katoh, Y. Swelling of nuclear graphite and high quality carbon fiber composite under very high irradiation temperature. *J. Nucl. Mater.* **2008**, *381*, 55–61. [CrossRef]
9. Zhao, Y.; Li, X.; Liu, C.; Yang, H.; Chen, B.; Qin, Y.; Xu, S.; Cheng, L.; Zhang, L. Irradiation effects on Amosic-3 silicon carbide composites by Si ions implantation. *J. Eur. Ceram. Soc.* **2019**, *39*, 4501–4509. [CrossRef]
10. Katoh, Y.; Ozawa, K.; Shih, C.; Nozawa, T.; Shinavski, R.J.; Hasegawa, A.; Snead, L.L. Continuous SiC fiber, CVI SiC matrix composites for nuclear applications: Properties and irradiation effects. *J. Nucl. Mater.* **2014**, *448*, 448–476. [CrossRef]
11. Agarwal, S.; Koyanagi, T.; Bhattacharya, A.; Wang, L.; Katoh, Y.; Hu, X.; Pagan, M.; Zinkle, S.J. Neutron irradiation-induced microstructure damage in ultra-high temperature ceramic TiC. *Acta Mater.* **2020**, *186*, 1–10. [CrossRef]
12. Yang, J.; Ye, F.; Cheng, L. In-situ synthesized nano-porous titanium carbide coating on silicon carbide fibres using titanium tetrachloride vapour. *Ceram. Int.* **2023**, *49*, 7996–8004. [CrossRef]
13. Barsoum, M.W. The $M_{N+1}AX_N$ phases: A new class of solids: Thermodynamically stable nanolaminates. *Prog. Solid State Chem.* **2000**, *28*, 201–281. [CrossRef]
14. Whittle, K.R.; Blackford, M.; Aughterson, R.; Moricca, S.; Lumpkin, G.R.; Riley, D.; Zaluzec, N. Radiation tolerance of $M_{n+1}AX_n$ phases, Ti_3AlC_2 and Ti_3SiC_2. *Acta Mater.* **2010**, *58*, 4362–4368. [CrossRef]
15. Liu, S.; Yang, T.; Zhang, J.; Yan, Z.; Lu, Y.; Han, D.; Wang, C.; Fang, Y.; Wang, Y. Thermal effects in ion irradiated Ti_2AlC and Ti_3SiC_2. *Nucl. Instrum. Methods Phys. Res. Sect. B Beam Interact. Mater. At.* **2018**, *435*, 50–55. [CrossRef]
16. Zhang, J.W.; Hu, C.F.; Wang, Y.G.; Huang, Q.; Cui, P. Interfacial reactions between polymer derived SiC fiber and $Ti_3Si(Al)C_2$. *Key Eng. Mater.* **2013**, *544*, 238–244. [CrossRef]
17. Yang, J.; Ye, F.; Cheng, L. In-situ formation of Ti_3SiC_2 interphase in SiC_f/SiC composites by molten salt synthesis. *J. Eur. Ceram. Soc.* **2022**, *42*, 1197–1207. [CrossRef]
18. Tao, P.; Liu, W.; Wang, Y. Fabrication of SiC_f/Ti_3SiC_2 composites with high thermal conductivity by spark plasma sintering. *Ceram. Int.* **2020**, *46*, 2571–2575. [CrossRef]
19. Cappia, F.; Cullison, M.; Chen, T.; Kombaiah, B.; Bawane, K.; Teng, F.; Madden, J.; Perez, E.; Yao, T.; Lei, P. Grain subdivision and structural modifications by high-energy heavy ions in UO_2 with different initial grain size. *Nucl. Instrum. Methods Phys. Res. Sect. B Beam Interact. Mater. At.* **2022**, *515*, 48–60. [CrossRef]
20. Yao, T.; Guo, X.; Lei, P.; Wang, Y.; Frankel, G.S.; Lian, J. Corrosion interactions between stainless steel and lead vanado-iodoapatite nuclear waste form part II. *NPJ Mater. Degrad.* **2020**, *4*, 15. [CrossRef]
21. Guo, X.; Gin, S.; Lei, P.; Yao, T.; Liu, H.; Schreiber, D.K.; Ngo, D.; Viswanathan, G.; Li, T.; Kim, S.H. Self-accelerated corrosion of nuclear waste forms at material interfaces. *Nat. Mater.* **2020**, *19*, 310–316. [CrossRef] [PubMed]
22. Guo, X.; Gin, S.; Lei, P.; Yao, T.; Liu, H.; Schreiber, D.K.; Ngo, D.; Viswanathan, G.; Li, T.; Kim, S.H. Reply to: How much does corrosion of nuclear waste matrices matter. *Nat. Mater.* **2020**, *19*, 962–963. [CrossRef] [PubMed]
23. Lei, P.; Yang, K.; Shi, T.; Wei, M.; Ran, G.; Lu, C. Surface alteration and chemical durability of hollandite (Cr, Al and Ti) consolidated by spark plasma sintering in acid solution. *J. Nucl. Mater.* **2022**, *568*, 153892. [CrossRef]
24. Lei, P.; Yao, T.; Gong, B.; Zhu, W.; Ran, G.; Lian, J. Spark plasma sintering-densified vanadinite apatite-based chlorine waste forms with high thermal stability and chlorine confinement. *J. Nucl. Mater.* **2020**, *528*, 151857. [CrossRef]
25. Lei, P.; Ji, X.; Qiu, J.; Si, J.; Peng, T.; Teng, C.; Wu, L. The Influence of Grain Size on Microstructure Evolution in CeO_2 under Xenon Ion Irradiation. *Nanomaterials* **2024**, *14*, 1498. [CrossRef]
26. Chen, J.; Ye, F.; Cheng, L.; Yang, J.; Chen, X. Preparation and properties of Ti_3SiC_2 preform reinforced SiC ceramic matrix composites. *J. Eur. Ceram. Soc.* **2023**, *43*, 3146–3157. [CrossRef]
27. Kimura, T. Molten salt synthesis of ceramic powders. In *Advances in Ceramics—Synthesis and Characterization, Processing and Specific Applications*; IntechOpen: London, UK, 2011; pp. 75–100.
28. Dash, A.; Sohn, Y.J.; Vaßen, R.; Guillon, O.; Gonzalez-Julian, J. Synthesis of Ti_3SiC_2 MAX phase powder by a molten salt shielded synthesis (MS3) method in air. *J. Eur. Ceram. Soc.* **2019**, *39*, 3651–3659. [CrossRef]
29. Ziegler, J.F.; Ziegler, M.D.; Biersack, J.P. SRIM–The stopping and range of ions in matter (2010). *Nucl. Instrum. Methods Phys. Res. Sect. B Beam Interact. Mater. At.* **2010**, *268*, 1818–1823. [CrossRef]
30. Emmerlich, J.; Music, D.; Eklund, P.; Wilhelmsson, O.; Jansson, U.; Schneider, J.M.; Högberg, H.; Hultman, L. Thermal stability of Ti_3SiC_2 thin films. *Acta Mater.* **2007**, *55*, 1479–1488. [CrossRef]
31. Qin, J.; He, D. Phase stability of Ti_3SiC_2 at high pressure and high temperature. *Ceram. Int.* **2013**, *39*, 9361–9367. [CrossRef]
32. Magnus, C.; Rainforth, W.M. Spark plasma sintering (SPS) synthesis and tribological behaviour of MAX phase composite of the family $Ti_{n+1}SiC_n$ (n = 2). *Wear* **2019**, *438*, 203062. [CrossRef]
33. He, G.; Xu, J.; Zhang, Z.; Qian, Y.; Zuo, J.; Li, M.; Liu, C. Interfacial reactions and mechanical properties of SiC fiber reinforced Ti_3SiC_2 and $Ti_3(SiAl)C_2$ composites. *Mater. Sci. Eng. A* **2021**, *827*, 142069. [CrossRef]
34. Barr, C.M.; Chen, E.Y.; Nathaniel, J.E.; Lu, P.; Adams, D.P.; Dingreville, R.; Boyce, B.L.; Hattar, K.; Medlin, D.L. Irradiation-induced grain boundary facet motion: In situ observations and atomic-scale mechanisms. *Sci. Adv.* **2022**, *8*, eabn0900. [CrossRef]
35. Zinkle, S. 1.03—Radiation-Induced effects on microstructure. *Compr. Nucl. Mater.* **2012**, *1*, 65–98.
36. Hobbs, L.W.; Clinard, F.W., Jr.; Zinkle, S.J.; Ewing, R.C. Radiation effects in ceramics. *J. Nucl. Mater.* **1994**, *216*, 291–321. [CrossRef]

37. Chen, C.-H.; Zhang, Y.; Fu, E.; Wang, Y.; Crespillo, M.L.; Liu, C.-Z.; Shannon, S.; Weber, W.J. Irradiation-induced microstructural change in helium-implanted single crystal and nano-engineered SiC. *J. Nucl. Mater.* **2014**, *453*, 280–286. [CrossRef]
38. Pramono, Y.; Sasaki, K.; Yano, T. Release and diffusion rate of helium in neutron-irradiated SiC. *J. Nucl. Sci. Technol.* **2004**, *41*, 751–755. [CrossRef]
39. Ye, C.; Xue, J.; Liu, T.; Shu, R.; Yan, Y; Liao, Y.; Ren, Q.; Ran, G.; Sun, K.; Jiang, L. The microstructure evolution in a SiC_f/SiC composite under triple ion beam irradiation. *Ceram. Int.* **2020**, *46*, 9901–9906. [CrossRef]
40. Liu, Y.-L.; Zhang, Y.; Zhou, H.-B.; Lu, G.-H.; Liu, F.; Luo, G.-N. Vacancy trapping mechanism for hydrogen bubble formation in metal. *Phys. Rev. B Condens. Matter Mater. Phys.* **2009**, *79*, 172103. [CrossRef]
41. Jastrzebski, Z.D.; Komanduri, R. *The Nature and Properties of Engineering Materials*; Wiley: Hoboken, NJ, USA, 1988.
42. Sun, J.; You, Y.-W.; Hou, J.; Li, X.; Li, B.; Liu, C.; Wang, Z. The effect of irradiation-induced point defects on energetics and kinetics of hydrogen in 3C-SiC in a fusion environment. *Nucl. Fusion* **2017**, *57*, 066031. [CrossRef]
43. Dong, W.; Shen, Q.; Wei, M.; Lei, P.; Song, L.; Chang, Q.; Ye, C. Research on the surface damage of Si^+ and H^+ co-implanted 6H-SiC before and after annealing. *Nucl. Instrum. Methods Phys. Res. Sect. B Beam Interact. Mater. At.* **2023**, *538*, 81–86. [CrossRef]

nanomaterials

Article

The Influence of Grain Size on Microstructure Evolution in CeO_2 under Xenon Ion Irradiation

Penghui Lei [1], Xiaoyu Ji [1], Jie Qiu [1,*], Jiaxuan Si [2], Tao Peng [2], Changqing Teng [2] and Lu Wu [2,*]

1 School of Nuclear Science and Technology, Xi'an Jiaotong University, Xi'an 710049, China; penghuilei@xjtu.edu.cn (P.L.); 2192112868@stu.xjtu.edu.cn (X.J.)
2 The First Sub-Institute, Nuclear Power Institute of China, Chengdu 610041, China; sijiaxuan9610@126.com (J.S.); pengtaoustb@163.com (T.P.); 21626040@zju.edu.cn (C.T.)
* Correspondence: qiu2021@xjtu.edu.cn (J.Q.); wulu1002@126.com (L.W.)

Abstract: Large-grained UO_2 is considered a potential accident-tolerant fuel (ATF) due to its superior fission gas retention capabilities. Irradiation experiments for cerium dioxide (CeO_2), used as a surrogate fuel, is a common approach for evaluating the performance of UO_2. In this work, spark plasma sintered CeO_2 pellets with varying grain sizes (145 nm, 353 nm, and 101 μm) and a relative density greater than 93.83% were irradiated with 4 MeV Xe ions at a fluence of 2×10^{15} ions/cm^2 at room temperature, followed by annealing at 600 °C for 3 h. Microstructure, including dislocation loops and bubble morphology of the irradiated samples, has been characterized. The average size of dislocation loops increases with increasing grain size. Large-sized dislocation loops are absent near the grain boundary because the boundary absorbs surrounding defects and prevents the dislocation loops from coalescing and expanding. The distribution of bubbles within the grain is uniform, whereas the large-sized and irregularly shaped xenon bubbles observed in the small grain exhibit pipe diffusion along the grain boundaries. The bubble diameter in the large-grained pellet is the smallest. As the grain size increases, the volumetric swelling of the irradiated pellets decreases while the areal density of Xe bubbles increases. Elemental segregation, which tends to occur at dislocation loops and grain boundaries, has been analyzed. Large-grained CeO_2 pellet with lower-density grain boundaries exhibits better resistance to volumetric swelling and elemental segregation, suggesting that large-grained UO_2 pellets could serve as a potential ATF.

Keywords: grain size; cerium dioxide; ion irradiation; dislocation loops; gas bubbles

1. Introduction

Irradiation-induced unconventional structural modification of UO_2 where fission reactions occur is termed high-burnup structure (HBS) [1–3]. These structures are characterized by the presence of nanocrystalline grains and micro-pores, which can lead to significant changes in fuel properties [4]. Changes in fuel properties can have a substantial impact on the economic efficiency and operational safety of nuclear power plants. Large-grained UO_2 is considered a potential ATF due to its superior fission gas retention capabilities and resistance to HBS formation, as well as its enhanced plasticity and thermal creep properties compared to conventional UO_2 [5,6]. Grain size influences numerous critical mechanisms that can lead to a decline in fuel performance [7]. Therefore, developing novel nanocrystalline UO_2 fuels that draw inspiration from the HBS is also a promising pathway [8,9]. Inert gases such as xenon, generated by fission reactions and possessing relatively low solubility in nuclear fuel, are easily trapped in grain boundaries or voids, leading to the formation of fission bubbles [10–12]. This process degrades the mechanical and thermal properties of the nuclear fuel, resulting in volume swelling, cracking, and decreased thermal conductivity [3]. Therefore, it is crucial to assess and understand the fission gas retention capabilities of UO_2 with different grain sizes.

CeO_2 is frequently used as a non-radioactive surrogate fuel for UO_2 to study microstructural evolution under irradiation due to their nearly identical lattice characteristics, fluorite structure, and similar physical properties (Table 1) [13–16]. Compared to traditional sintering techniques, Spark Plasma Sintering (SPS) enables the sintering of powder samples into dense pellets without the need for additives at significantly lower temperatures and shorter duration [17–21], which means the growth of grains in the sintered progress is greatly inhibited by SPS. Fission gas migration in fuel is one of the most critical parameters for the economic efficiency and safety of nuclear fuels. Dose rate and temperature can significantly influence the dislocation loop characteristics in Kr^+-irradiated CeO_2 during in-situ experiments [15]. According to Yasuda et al. [22], 210 MeV Xe ion irradiation to CeO_2 induces high-density electronic excitations, leading to microstructural modifications such as dot-contrast features, dislocation loops, line dislocations, and the formation of subdivided grains. CeO_2 thin films exposed to 92 MeV Xe^{23+} radiation were found to undergo significant microstructural rearrangements and the formation of secondary phases as a result of the high-energy ion irradiation [23]. Xia et al. [24] demonstrated through molecular dynamics and static simulations that the diffusion activation energy and formation energies of Xe atoms at grain boundaries in UO_2 are significantly lower than those in the bulk. Numerous studies have been published on the impact of grain boundaries on gas migration properties [25–28], but there is limited information available on the elemental distribution and segregation in SPS-densified pellets with varying grain sizes.

Table 1. Summary of the similar properties of UO_2 and CeO_2 pellets.

ID	Melting Point (°C)	Fluorite Structure	Thermal Diffusivity	microhardness (GPa)
UO_2	2870	Yes	Up to 700 °C	6.8–12.9 [29]
CeO_2	2600	Yes	Up to 700 °C	6.38–7.4 (This work)

In this work, SPS fabrication was employed to sinter three types of CeO_2 pellets with different grain sizes. In addition to phase stability, potential CeO_2 pellets with varying grain sizes should also possess desirable mechanical properties to enhance performance. The effects of grain size on resistance to volumetric swelling, elemental segregation, and dislocation loop formation have been revealed.

2. Material and Methods

2.1. Preparation of CeO_2 with Controlled Microstructure

The dense CeO_2 ceramics with different grain sizes and microstructures were sintered by using SPS technique. Polycrystalline CeO_2 pellets with different grain sizes (145 nm, 353 nm, and 101 μm) were fabricated by SPS from CeO_2 powders (99.95% purity, <50 nm, Shanghai Aladdin Biochemical Technology Co., LTD, Shanghai, China). A total of 1.3 g of CeO_2 powder was loaded into a die with an inner diameter of 10 mm for each sintering process. Nanocrystalline CeO_2 pellets, with a grain size of 145 nm, were sintered in a tungsten carbide (WC) die for 5 min at 750 °C under a pressure of 120 MPa. The pellets, with a grain size of 353 nm, were sintered in a tungsten carbide (WC) die for 5 min at 1000 °C under a pressure of 120 MPa. The control temperature in these two sintering routes was monitored by a K-type thermocouple. Large-sized CeO_2 pellets with a grain size of 101 μm were sintered at 1550 °C for 10 min under 40 MPa. To avoid potential interaction between graphite and CeO_2 during the sintering process, boron nitride spray was applied to the graphite foil wrapped around the inner wall of the die before loading the CeO_2 powders. A pyrometer was used to monitor sintering temperature above 1000 °C. For all the sintering, the temperature increase rate was set at 100 °C per minute, and an initial pressure of 10 MPa was applied and then linearly increased to the peak pressure with temperatures. The bulk density of the sintered pellets was measured based on Archimedes principle by using an immersing method with deionized water at room temperature (RT) with an Ohaus PX124ZH density profiler (OHAUS Corporation, Pine Brook, NJ, USA).

2.2. Irradiation Experiments

The as-sintered CeO_2 pellets were ground from 400 grid to 5000 grid by using SiC sandpapers, and then polished with 1 μm diamond suspensions. The polished samples were irradiated at room temperature by 4 MeV Xe ions with a fluence of 2×10^{15} ions/cm^2 using the 320 kV platform for multi-discipline research with highly charged ions at the Institute of Modern Physics, CAS (Lanzhou, China). The irradiation damage profile and the depth distribution of implanted Xe ions were calculated using the quick Kinchin–Pease model with the Stopping and Range of Ions in Matter (SRIM) 2013 software [30,31]. The threshold displacement energies for Ce and O elements in cerium dioxide were set to 56 eV and 27 eV, respectively, as reported in reference [32]. The concentration of Xe ions reaches a peak value of approximately 0.05% at a depth of 880 nm, with a displacement damage of 3.35 dpa, as shown in Figure 1. Post-irradiation annealing was performed in a muffle furnace placed in a glovebox. The annealing temperature was set to 600 °C for 3 h to simulate the edge temperature of the pellets during service.

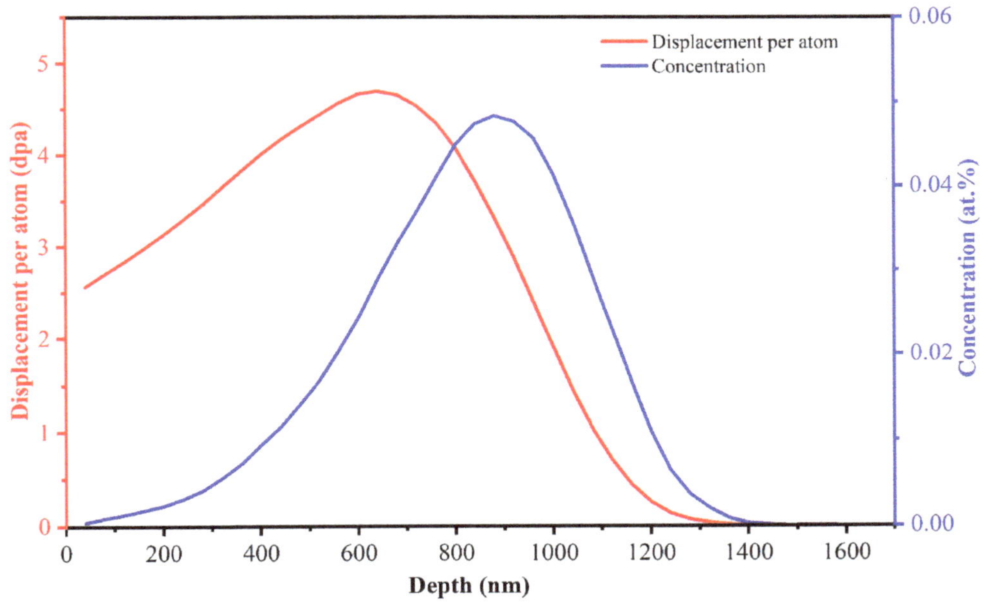

Figure 1. The damage profile and the concentration distribution of implanted Xe ions in CeO_2 with 4 MeV Xe ions at a fluence of 2×10^{15} ions/cm^2 calculated by SRIM 2013.

2.3. Microstructural Characterizations

XRD was used to study the phase constitution of the sintered pellets. A Gemini SEM 500 SEM (Carl Zeiss AG, Oberkochen, Germany) was used to observe and analyze the fracture surfaces of the samples. Microstructure and morphology characterization using TEM were conducted with a Talos F200X. The cross-sectional TEM samples of the post-irradiation annealed pellet were prepared by focused ion beam (FIB) using a Thermo Scientific Helios 5 UX (Thermo Fisher Scientific, Waltham, MA, USA), achieving a uniform thickness of approximately 100 nm. Micro-indentation testing was carried out to acquire the Vickers microhardness of the sintered samples. Indentation was performed using a diamond-shaped indenter on the polished surface with a load of 1.0 kgf (~9.8N) for 10 s holding, and post indentation measurement was carried out by using SEM. The general approach to estimate the Vickers hardness values is based on the following equation [33]:

$$H = 1.8544\left(\frac{P}{\alpha^2}\right)\left[\text{kgf}/\text{mm}^2\right]$$

where H is the values of estimated Vickers hardness, α is the average diagonal length of the indentation, and P is the applied load.

3. Results and Discussion

3.1. Microstructure Morphology and Mechanical Properties of the Sintered Samples

The XRD patterns of as-sintered pellets, from 750 °C to 1550 °C, display characteristic diffraction peaks that are well indexed to CeO_2 as shown in Figure 2. The results indicate that the phase constitution of the sintered pellets by SPS is single-phase CeO_2, and the stoichiometry of the CeO_2 sintered pellets remains constant during different SPS sintering temperatures.

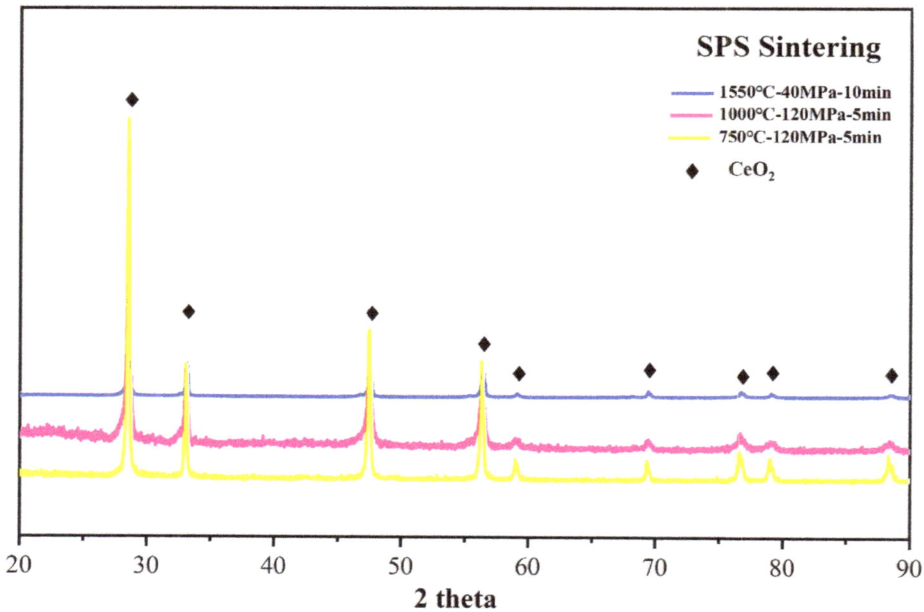

Figure 2. X-ray diffraction patterns of the sintered CeO_2 pellets from 750 °C to 1550 °C.

Figure 3a illustrates the cross-sectional fractured surface of the sample sintered at 750 °C for 5 min. It can be observed that the sample contains both small-sized closed pores (indicated by red arrows) and large open pores (indicated by blue arrows), which are distributed around the grain boundaries. The presence of these large open pores inhibits the migration of grain boundaries, thereby decelerating the growth of grains at the same sintering temperature. Moreover, the presence of small-sized closed holes at the triple junction points of grains will serve to reduce the grain coarsening kinetics and restrict grain boundary migration [34,35]. The average grain size of the pellet sintered at 750 °C is 145 nm measured from the SEM image, with a theoretical density of 94.16%. The microstructure of the cross-sectional fractured surface of the sample sintered at 1000 °C for 5 min exhibits similar characteristics, with a much denser theoretical density of 96.15%, as observed in Figure 3b. The average grain size is 353 nm. The occurrence of both trans-granular (indicated by red circle) and intergranular (indicated by blue circle) fracture mechanisms is observed in Figure 3b, and the preferential propagation of cracks is along the weaker grain boundaries [36]. Upon increasing the sintering temperature to 1550 °C for 10 min, significant grain growth is observed, with an average grain size increasing to 101 µm and a theoretical density of 93.83%. Despite the complete elimination of large-sized open pores around grain boundaries, small-sized closed pores (indicated by red arrows) embedded within the grains can still be observed on the fracture surfaces. The fracture surface is characterized by an intergranular fracture model.

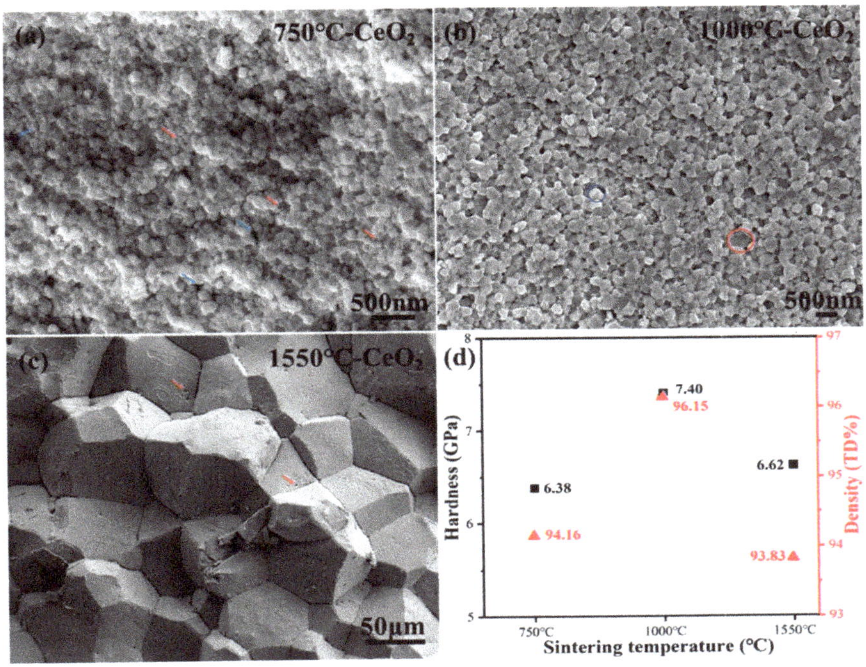

Figure 3. The cross-section fracture surfaces of the sintered CeO_2 pellets at different sintering temperatures (**a**) 750 °C; (**b**) 1000 °C; (**c**)1550 °C; and Vickers hardness and density of as-sintered CeO_2 pellets (**d**).

The microstructure characteristics of the sintered ceramic samples, particularly the grain sizes and densities, significantly impact their mechanical properties. Since the sintering process involves both the coarsening of the grains and the densification of the material through pore removal, these two variables are interrelated [37]. Grain size and density are found to be highly correlated with the hardness values of CeO_2 pellets sintered by SPS. As displayed in Figure 3d, the measured hardness initially increases with grain size up to 353 nm and subsequently exhibits a reduction in hardness with further increases in grain size. The pellet sintered at 1000 °C exhibited the highest density (96.15%) and the maximum hardness value (7.40 GPa) among the samples tested. For the pellets sintered at 750 °C and 1550 °C, pores and micro-cracks around the grain boundaries (Figure 3a,c) negatively impact the mechanical integrity. Consequently, much lower hardness values are observed. These results suggest that the elastic and plastic behaviors of as-sintered CeO_2 pellets upon indentation are mostly determined by grain size and density.

3.2. Microstructure of the Irradiated Samples

STEM/TEM images showing the microstructure of CeO_2 pellets irradiated with 4 MeV Xe ions with a fluence of 2×10^{15} ions/cm^2 at room temperature followed by annealing at 600 °C for 3 h are displayed in Figure 4. exhibits. The incident direction of Xe ions and the peak concentration depth are indicated in the STEM bright field image (Figure 4a) of the pellet sintered at 750 °C. It is evident from the STEM image that these grains in the irradiation area have not undergone amorphization. A grain of approximately 180 nm in size, marked by the red dashed box A in Figure 4a, is subjected to comprehensive analysis in Figure 4b. The grain is situated at a depth of 880 nm below the irradiated surface, with a displacement damage of 3.35 dpa and a peak Xe ion concentration of approximately 0.05%. The selected area electron diffraction (SAED) pattern from the grain is shown in the inserted image in the upper right corner of Figure 4b. The diffraction spots nearest the central spot originate from the (0 2 −2) and (−2 0 −2) crystal planes, with the diffraction zone axis being [−1 1 1], as shown in the insert. The STEM bright field image in Figure 4b, which was taken under the diffraction condition of the [−1 1 1] zone axis, shows the morphology and distribution of dislocation loops in the CeO_2 after irradiation followed by annealing at 600 °C for 3 h. The average size of dislocation loops is 3.2 nm. Figure 4c presents a high-resolution TEM (HRTEM) image of the peak concentration damage area of the grain for further visualization and structural verification. The atomic-resolution TEM image exhibits a sharp lattice image along the [−1 1 1] zone axis. The measured d-spacings are 0.194 nm and 0.192 nm, which are consistent with the PDF# 34-0394-CeO_2 (d = 0.191nm) in the Jade database, respectively. Meanwhile, atomic column misalignment regions are observed in the blue dashed box, indicating the presence of dislocation loops and defects caused by irradiation.

The STEM bright field image of the pellet sintered at 1000 °C shows the direction of Xe ions' incidence and the depth of the peak concentration in Figure 4d. Figure 4e illustrates the results of a grain measuring approximately 600 nm in diameter, as indicated by the red dashed box B in Figure 4d. As indicated in the inset of Figure 4e, the diffraction zone axis of the SAED pattern from the grain is [−1 1 1], and the diffraction spots closest to the center spot originate from the (0 2 −2) and (−2 0 −2) crystal planes. The diffraction condition of the [−1 1 1] zone axis is used to capture the STEM dark field image in Figure 4e, which displays the typical features of dislocation loops with a bimodal size distribution in the CeO_2 following irradiation and a 3-h annealing process at 600 °C. The dislocation loops have an average diameter of 5.56 nm. The presence of large dislocation loops, with diameters reaching up to 11.80 nm, alongside smaller loops (3.07 nm), suggests that dislocation loop nucleation and growth occurred simultaneously. Large-sized dislocation loops are absent near the grain boundary because the boundary absorbs surrounding defects and prevents the dislocation loops from coalescing and expanding. Grain boundaries have the capacity to function as neutral sinks, whereby they can remove defects through absorption and subsequent recombination of interstitials and vacancies [38]. For additional visualization and structural validation, an HRTEM image of the highest ion concentration region along the [−1 1 1] zone axis of the grain is shown in Figure 4f. The d-spacings of PDF# 34-0394-CeO_2 (d = 0.191 nm) are in line with the measured d-spacings of 0.185 nm and 0.187 nm, respectively. In the meanwhile, the blue dashed box shows a misalignment of the atomic column.

For the pellets sintered at 1550 °C, microstructure morphologies are shown in Figure 4g. Figure 4h displays the results of a thorough investigation as indicated by the red dashed box C in Figure 4g. The SAED pattern, shown in the inset of Figure 4h, indicates that the diffraction zone axis of the grain is [−1 1 1]. The diffraction condition of the [−1 1 1] zone axis was used to obtain the STEM dark field image shown in Figure 4h, which highlights the characteristic features of dislocation loops in the CeO_2. The average diameter of the dislocation loops is 10.87 nm. Figure 4i presents an HRTEM image of the region with the highest concentration of Xe ions, oriented along the [1 1 1] zone axis of the grain. The measured d-spacings of 0.189 nm and 0.192 nm are in line with the d-spacings of the PDF#

34-0394-CeO_2 (d = 0.191 nm), respectively. Meanwhile, atomic column misalignment is indicated by the blue dashed box. It can be concluded that the average size of dislocation loops increases with increasing grain size, consistent with a previous study [39] showing the dislocation loop size in grade 91 steel subjected to Fe ion irradiation at 300 °C also increases with grain size.

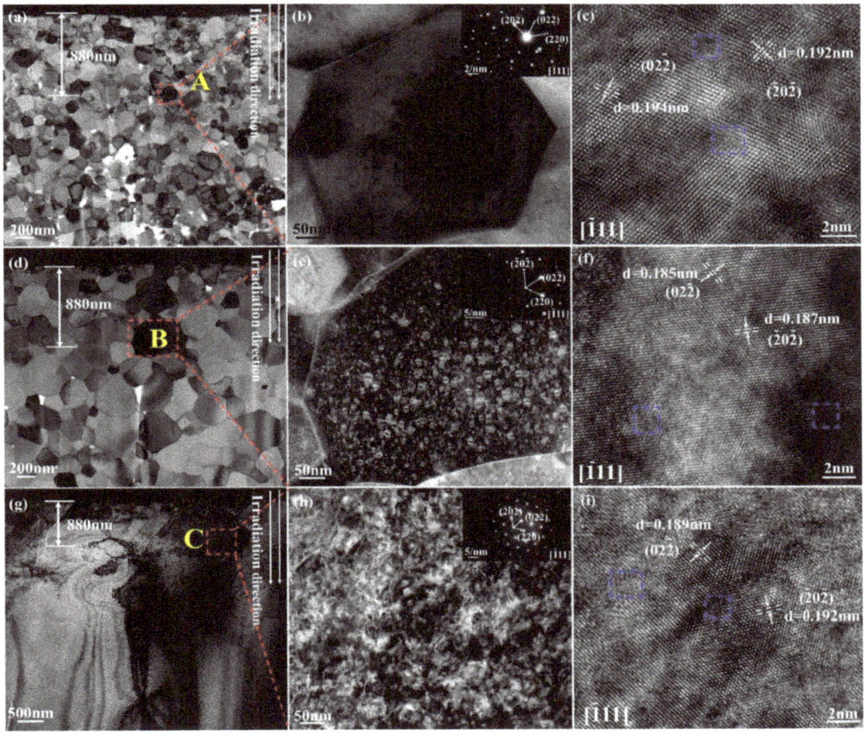

Figure 4. The microstructure of CeO_2 pellets irradiated with 4 MeV Xe ions with a fluence of 2×10^{15} ions/cm^2 at room temperature followed by annealing at 600 °C for 3 h: (**a**,**c**) the pellet sintered at 750 °C; (**d**–**f**) the pellet sintered at 1000 °C; (**g**–**i**) the pellet sintered at 1550 °C; the SAED patterns of CeO_2 pellets were inserted in the images (**b**,**e**,**h**).

3.3. *Evolution of Gas Bubbles and Element Migration Behavior*

The distribution of Xe gas bubbles in the 4 MeV irradiated CeO_2 pellets, followed by a 3 h annealing at 600 °C, is also investigated with the bubble size and density quantitatively analyzed in Figure 5. The nano-sized bubbles in the sample sintered at 750 °C for 5 min captured using under-focused bright field TEM imaging are illustrated in Figure 5a. It can be observed that the bubbles diffuse along the grain boundaries through pipe diffusion. Xe gas bubbles are uniformly distributed within the grain, with an average diameter of 2.10 ± 0.04 nm. For the pellet sintered at 1000 °C, bubbles gather along the grain boundaries. The distribution of bubbles within the grain is uniform, with an average size of approximately 2.02 ± 0.04 nm. Grain boundaries are prone to trapping bubbles due to their role as defect sinks [25]. The large-sized and irregularly shaped xenon bubbles observed along the grain boundaries in Figure 5a,b are attributed to the pipe diffusion mechanism. This mechanism explains how bubbles grow along grain boundaries, forming coarse and irregular shapes due to the low activation energy and high mobility of the diffusion process [40]. High-density smaller bubbles can be observed through an under-focused

TEM image (Figure 5c) showing the morphology and distribution of Xe bubbles. Bubbles are uniformly distributed within the grain, with an average diameter of approximately 1.31 ± 0.03 nm.

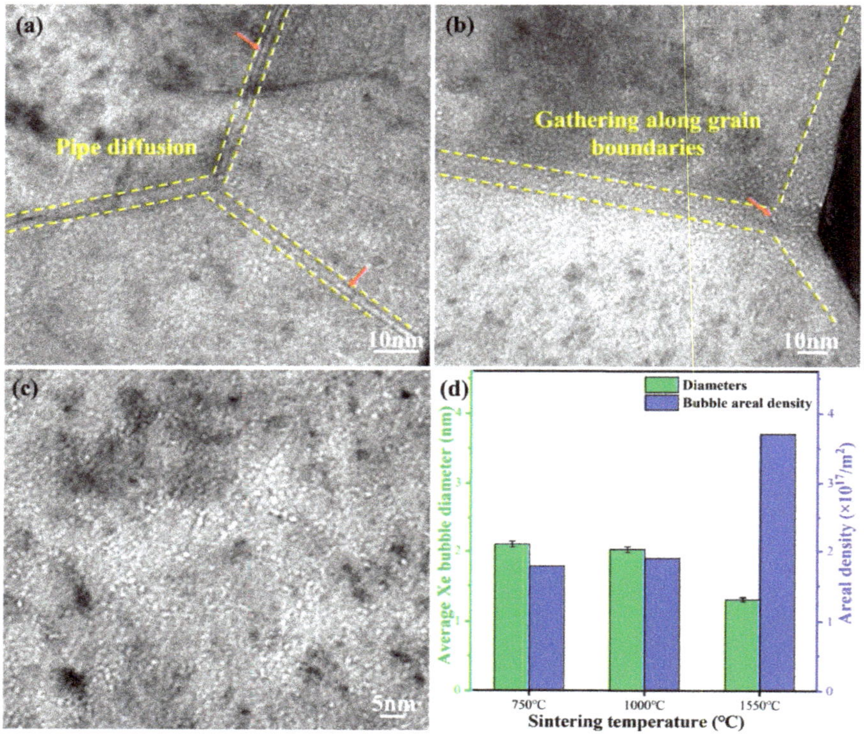

Figure 5. The distribution of Xe gas bubbles in the 4 MeV irradiated CeO_2 pellets followed by annealing at 600 °C for 3 h: (**a**) the pellet sintered at 750 °C; (**b**) the pellet sintered at 1000 °C; (**c**) the pellet sintered at 1550 °C; and (**d**) average diameters and areal density of Xe bubbles.

Figure 5d presents a histogram describing the bubble diameter and areal densities based on measurements taken across grains of different sizes (Table 2). There is a significant shift in the areal density of Xe bubbles, increasing from an initial $1.78 \times 10^{17}/m^2$ to a final $3.70 \times 10^{17}/m^2$, corresponding with increasing grain size. The volumetric swelling rate induced by Xenon bubbles is calculated using the equation provided in reference [41], assuming a thickness of 100 nm for the FIB lamella. The estimated volumetric swelling slightly decreases with increasing grain size, from 0.95% at 145 nm to 0.90% at 353 nm, and then further decreases to 0.49% at 101 μm. The present findings confirm the enhanced resistance to gas precipitation and volumetric swelling in large-grain CeO_2, in line with the previous report on UO_2 [26]. The Xe diffusivities at uranium dioxide grain boundaries, as derived from molecular dynamics simulations, exceed those in the bulk by more than six orders of magnitude [27]. Miao et al. highlighted that the excessive dislocations and high density of grain boundaries in UO_2 may enhance the diffusion of both Xe atoms and vacancies, thereby promoting the formation of micropores [28], which could ultimately result in the cracking of fuel pellets. In large-sized grains, the migration path of fission gas to grain boundaries is longer, making it more difficult for bubbles to diffuse and aggregate. In summary, the formation of small-sized bubbles with a uniform distribution suggests that large-sized CeO_2 exhibits favorable resistance to volumetric swelling.

Table 2. Summary of sintering conditions, average Xe bubble diameters, areal density of Xe bubbles, and estimated swelling of CeO_2 pellets.

Sintering Temp (°C)	Grain Size	Average Xe Bubble Diameter	Areal Density of Xe Bubble ($\times 10^{17}/m^2$)	Estimated Swelling
750	145 nm	2.1	1.78	0.95%
1000	353 nm	2.02	1.89	0.90%
1550	101 μm	1.31	3.7	0.49%

Irradiation-induced elemental segregation (RIS) at dislocation loops in the 4 MeV irradiated CeO_2 pellets followed by 600 °C annealing for 3 h was examined using STEM-EDS (energy dispersive spectrometer) analysis. Figure 6 (a) presents a STEM dark-filed image of dislocation loops under the [−1 1 1] zone axis in irradiated CeO_2 pellets sintered at 1000 °C. The dislocation loops exhibit a bimodal size distribution, with the diameters of the larger loops reaching up to 11.80 nm. To gain more insights into the dislocation loop, EDS line scans were employed to analyze elemental segregation, as shown in Figure 6b. The dislocation loop marked by the blue arrow in Figure 6a indicates that Ce and Xe tend to accumulate around the edge of the loop, while O is depleted in that region. As reported by Cao et al. [14], the dislocation loop in CeO_2 is considered a strong sink for absorbing point defects and defect clusters induced by irradiation. The observed elemental segregation around the edge of the loop is attributed to the absorption activity of the dislocation loop. Meanwhile, Xe bubbles have accumulated along the grain boundaries, indicating that xenon segregation has also occurred in these regions. Modifications to the chemical composition and atomic structure of grain boundaries can lead to significant changes in a material's corrosion resistance, radiation tolerance, and imperviousness to fission products [42]. Therefore, large-sized CeO_2 pellet with lower-density grain boundaries exhibits more favorable resistance to corrosion and elemental segregation.

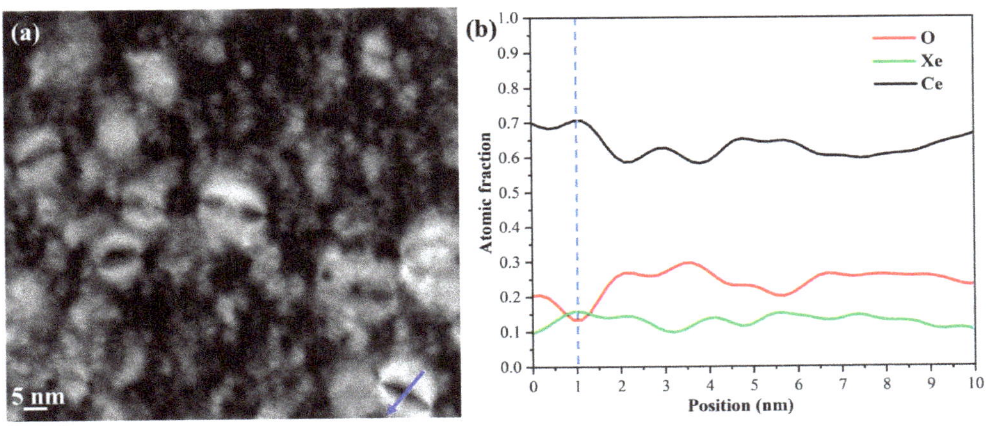

Figure 6. (**a**) A STEM dark filed image showing dislocation loops in irradiated CeO_2 pellet sintered at 1000 °C; and (**b**) EDS line scan along the blue arrow across the dislocation loop in (**a**).

4. Conclusions

In this paper, SPS sintering is applied to fabricate polycrystalline CeO_2 pellets with different grain sizes (145 nm, 353 nm, and 101 μm) as a surrogate fuel for UO_2. Radiation-induced microstructure evolution and bubble behavior of CeO_2 pellets irradiated by 4 MeV Xe ions at room temperature followed by annealing at 600 °C for 3 h are studied. Atomic column misalignment regions are observed in the HRTEM images, indicating the presence of dislocation loops and defects caused by irradiation. The average size of dislocation loops under the [−1 1 1] zone axis increases with increasing grain size of CeO_2 pellets. The large-sized and irregularly shaped xenon bubbles due to the pipe diffusion mechanism are observed along the grain boundaries in small-sized grains. Bubbles are uniformly distributed within the large-sized grain (101 μm), with an average diameter of approximately 1.31 ± 0.03 nm, the smallest observed in the study. There is a significant shift in the areal density of Xe bubbles, increasing from an initial $1.78 \times 10^{17}/m^2$ to a final $3.70 \times 10^{17}/m^2$, corresponding with increasing grain size. The estimated volumetric swelling slightly decreases with increasing grain size, from 0.95% at 145 nm to 0.90% at 353 nm, and then further decreases to 0.49% at 101 μm. Irradiation-induced elemental segregation has occurred around the edge of the dislocation loop and the grain boundary. This work provides a further understanding of how large-sized CeO_2/UO_2 pellets exhibit more favorable resistance to volumetric swelling, corrosion, and elemental segregation.

Author Contributions: Conceptualization, P.L. and J.Q.; methodology, X.J.; software, J.S.; formal analysis, T.P.; investigation, C.T.; writing—original draft preparation, P.L.; writing—review and editing, J.Q. and L.W.; funding acquisition, L.W. All authors have read and agreed to the published version of the manuscript.

Funding: This work is supported by the National Key Research and Development Program of China (No. 2022YFB1902401), the National Natural Science Foundation of China (12205232), and the Basic Scientific Research Fund of Xi'an Jiaotong University (xzy012022077).

Data Availability Statement: The data presented in this study are available on request from the corresponding authors.

Conflicts of Interest: The authors declare no conflicts of interest.

References

1. Matzke, H. On the rim effect in high burnup UO_2LWR fuels. *J. Nucl. Mater.* **1992**, *189*, 141–148. [CrossRef]
2. Xiao, H.; Long, C.; Chen, H. The formation mechanisms of high burnup structure in UO_2 fuel. *J. Nucl. Mater.* **2021**, *556*, 153151. [CrossRef]
3. Jiang, Y.; La, Y.; Liu, X.; Liu, W. Effect of grain size on irradiation-induced bubble evolution in high burn-up UO_2: A phase-field study. *J. Nucl. Mater.* **2024**, *601*, 155312. [CrossRef]
4. Ronchi, C.; Sheindlin, M.; Staicu, D.; Kinoshita, M. Effect of burn-up on the thermal conductivity of uranium dioxide up to 100.000 MWd t − 1. *J. Nucl. Mater.* **2004**, *327*, 58–76. [CrossRef]
5. Kang, K.W.; Yang, J.H.; Kim, J.H.; Rhee, Y.W.; Kim, D.J.; Kim, K.S.; Song, K.W. Effects of MnO-Al_2O_3 on the grain growth and high-temperature deformation strain of UO_2 fuel pellets. *J. Nucl. Sci. Technol.* **2010**, *47*, 304–307. [CrossRef]
6. Zhong, Y.; Gao, R.; Li, B.; Yang, Z.; Huang, Q.; Wang, Z.; Duan, L.; Liu, X.; Chu, M.; Zhang, P. Preparation and characterization of large grain UO_2 for accident tolerant fuel. *Front. Mater.* **2021**, *8*, 651074. [CrossRef]
7. Yu, Z.; Xu, X.; Chen, W.-Y.; Sharma, Y.; Wang, X.; Chen, A.; Ulmer, C.J.; Motta, A.T. In-situ irradiation-induced studies of grain growth kinetics of nanocrystalline UO_2. *Acta Mater.* **2022**, *231*, 117856. [CrossRef]
8. Yao, T.; Mo, K.; Yun, D.; Nanda, S.; Yacout, A.M.; Lian, J. Grain growth and pore coarsening in dense nano-crystalline UO_{2+x} fuel pellets. *J. Am. Ceram. Soc.* **2017**, *100*, 2651–2658. [CrossRef]
9. Miao, Y.; Yao, T.; Lian, J.; Park, J.-S.; Almer, J.; Bhattacharya, S.; Yacout, A.M.; Mo, K. In situ synchrotron investigation of grain growth behavior of nano-grained UO_2. *Scr. Mater.* **2017**, *131*, 29–32. [CrossRef]
10. Zacharie, I.; Lansiart, S.; Combette, P.; Trotabas, M.; Coster, M.; Groos, M. Thermal treatment of uranium oxide irradiated in pressurized water reactor: Swelling and release of fission gases. *J. Nucl. Mater.* **1998**, *255*, 85–91. [CrossRef]
11. Baker, C. The migration of intragranular fission gas bubbles in irradiated uranium dioxide. *J. Nucl. Mater.* **1977**, *71*, 117–123. [CrossRef]
12. Fisher, S.; White, R.; Cook, P.; Bremier, S.; Corcoran, R.; Stratton, R.; Walker, C.T.; Ivison, P.; Palmer, I. Microstructure of irradiated SBR MOX fuel and its relationship to fission gas release. *J. Nucl. Mater.* **2002**, *306*, 153–172. [CrossRef]
13. Ye, B.; Oaks, A.; Kirk, M.; Yun, D.; Chen, W.-Y.; Holtzman, B.; Stubbins, J.F. Irradiation effects in UO_2 and CeO_2. *J. Nucl. Mater.* **2013**, *441*, 525–529. [CrossRef]
14. Cao, Z.; He, K.; Ran, G.; Qiu, X.; Sun, D.; Li, Y.; Xin, Y. Influence of thermal effect on dislocation loop evolution in Fe^+-irradiated CeO_2 during in-situ annealing. *Ceram. Int.* **2022**, *48*, 32022–32026. [CrossRef]
15. Yuan, F.; Cao, Z.; Cui, D.; Xin, Y.; Li, Y.; Sun, D.; Ran, G. Understanding the dose-rate effect on loop characteristics in Kr^+-irradiated CeO_2 by in-situ TEM study. *Ceram. Int.* **2024**, *50*, 28573–28583. [CrossRef]

16. Sonoda, T.; Kinoshita, M.; Chimi, Y.; Ishikawa, N.; Sataka, M.; Iwase, A. Electronic excitation effects in CeO_2 under irradiations with high-energy ions of typical fission products. *Nucl. Instrum. Methods Phys. Res. Sect. B Beam Interact. Mater. At.* **2006**, *250*, 254–258. [CrossRef]
17. Gong, B.; Yao, T.; Lei, P.; Harp, J.; Nelson, A.T.; Lian, J. Spark plasma sintering (SPS) densified U_3Si_2 pellets: Microstructure control and enhanced mechanical and oxidation properties. *J. Alloys Compd.* **2020**, *825*, 154022. [CrossRef]
18. Gong, B.; Yao, T.; Lei, P.; Cai, L.; Metzger, K.E.; Lahoda, E.J.; Boylan, F.A.; Mohamad, A.; Harp, J.; Nelson, A.T. U_3Si_2 and UO_2 composites densified by spark plasma sintering for accident-tolerant fuels. *J. Nucl. Mater.* **2020**, *534*, 152147. [CrossRef]
19. Gong, B.; Cai, L.; Lei, P.; Metzger, K.E.; Lahoda, E.J.; Boylan, F.A.; Yang, K.; Fay, J.; Harp, J.; Lian, J. Cr-doped U_3Si_2 composite fuels under steam corrosion. *Corros. Sci.* **2020**, *177*, 109001. [CrossRef]
20. Yao, T.; Gong, B.; Lei, P.; Lu, C.; Xu, P.; Lahoda, E.; Lian, J. UO_2 + 5 vol% ZrB_2 nano composite nuclear fuels with full boron retention and enhanced oxidation resistance. *Ceram. Int.* **2020**, *46*, 26486–26491. [CrossRef]
21. Guo, X.; Gin, S.; Lei, P.; Yao, T.; Liu, H.; Schreiber, D.K.; Ngo, D.; Viswanathan, G.; Li, T.; Kim, S.H. Self-accelerated corrosion of nuclear waste forms at material interfaces. *Nat. Mater.* **2020**, *19*, 310–316. [CrossRef] [PubMed]
22. Yasuda, K.; Etoh, M.; Sawada, K.; Yamamoto, T.; Yasunaga, K.; Matsumura, S.; Ishikawa, N. Defect formation and accumulation in CeO_2 irradiated with swift heavy ions. *Nucl. Instrum. Methods Phys. Res. Sect. B Beam Interact. Mater. At.* **2013**, *314*, 185–190. [CrossRef]
23. Popel, A.; Le Solliec, S.; Lampronti, G.; Day, J.; Petrov, P.; Farnan, I. The effect of fission-energy Xe ion irradiation on the structural integrity and dissolution of the CeO_2 matrix. *J. Nucl. Mater.* **2017**, *484*, 332–338. [CrossRef]
24. Xia, Y.; Wang, Z.; Wang, L.; Chen, Y.; Liu, Z.; Wang, Q.; Wu, L.; Deng, H. Molecular Dynamics Simulations of Xe Behaviors at the Grain Boundary in UO_2. *Metals* **2022**, *12*, 763. [CrossRef]
25. Yan, X.; Li, Z.; Lei, P.; Wang, S.; Gao, R. Significant suppression of helium bubbles in oxide dispersion strengthened FeCrAl alloys irradiated by high concentration of helium. *Vacuum* **2024**, *221*, 112915. [CrossRef]
26. Cappia, F.; Cullison, M.; Chen, T.; Kombaiah, B.; Bawane, K.; Teng, F.; Madden, J.; Perez, E.; Yao, T.; Lei, P. Grain subdivision and structural modifications by high-energy heavy ions in UO_2 with different initial grain size. *Nucl. Instrum. Methods Phys. Res. Sect. B Beam Interact. Mater. At.* **2022**, *515*, 48–60. [CrossRef]
27. Liu, X.-Y.; Galvin, C.O.; Cooper, M.W.D.; Andersson, A.D.R. *Molecular Dynamics Simulations of Fission Gas Xenon (Xe) Diffusion at UO_2 Grain-Boundaries (Rev. 1)*; Los Alamos National Laboratory (LANL): Los Alamos, NM, USA, 2023.
28. Miao, Y.; Yao, T.; Lian, J.; Zhu, S.; Bhattacharya, S.; Oaks, A.; Yacout, A.M.; Mo, K. Nano-crystallization induced by high-energy heavy ion irradiation in UO_2. *Scr. Mater.* **2018**, *155*, 169–174. [CrossRef]
29. Gong, B.; Frazer, D.; Yao, T.; Hosemann, P.; Tonks, M.; Lian, J. Nano-and micro-indentation testing of sintered UO_2 fuel pellets with controlled microstructure and stoichiometry. *J. Nucl. Mater.* **2019**, *516*, 169–177. [CrossRef]
30. Iwamoto, Y.; Meigo, S.-I.; Hashimoto, S. Estimation of reliable displacements-per-atom based on athermal-recombination-corrected model in radiation environments at nuclear fission, fusion, and accelerator facilities. *J. Nucl. Mater.* **2020**, *538*, 152261. [CrossRef]
31. Nordlund, K.; Zinkle, S.J.; Sand, A.E.; Granberg, F.; Averback, R.S.; Stoller, R.E.; Suzudo, T.; Malerba, L.; Banhart, F.; Weber, W.J. Primary radiation damage: A review of current understanding and models. *J. Nucl. Mater.* **2018**, *512*, 450–479. [CrossRef]
32. Guglielmetti, A.; Chartier, A.; van Brutzel, L.; Crocombette, J.-P.; Yasuda, K.; Meis, C.; Matsumura, S. Atomistic simulation of point defects behavior in ceria. *Nucl. Instrum. Methods Phys. Res. Sect. B Beam Interact. Mater. At.* **2008**, *266*, 5120–5125. [CrossRef]
33. *ASTM C1327-15*; Standard Test Method for Vickers Indentation Hardness of Advanced Ceramics. ASTM International: West Conshohocken, PA, USA, 2019.
34. Liao, S.C.; Mayo, W.E.; Pae, K.D. Theory of high pressure/low temperature sintering of bulk nanocrystalline TiO_2. *Acta Mater.* **1997**, *45*, 4027–4040. [CrossRef]
35. Lei, P.; Yao, T.; Gong, B.; Zhu, W.; Ran, G.; Lian, J. Spark plasma sintering-densified vanadinite apatite-based chlorine waste forms with high thermal stability and chlorine confinement. *J. Nucl. Mater.* **2020**, *528*, 151857. [CrossRef]
36. Yao, T.; Scott, S.; Xin, G.; Lu, F.; Lian, J. Dense Iodoapatite Ceramics Consolidated by Low-Temperature Spark Plasma Sintering. *J. Am. Ceram. Soc.* **2015**, *98*, 3733–3739. [CrossRef]
37. Lei, P.; Yang, K.; Shi, T.; Wei, M.; Ran, G.; Lu, C. Surface alteration and chemical durability of hollandite (Cr, Al and Ti) consolidated by spark plasma sintering in acid solution. *J. Nucl. Mater.* **2022**, *568*, 153892. [CrossRef]
38. El-Atwani, O.; Esquivel, E.; Efe, M.; Aydogan, E.; Wang, Y.; Martinez, E.; Maloy, S.A. Loop and void damage during heavy ion irradiation on nanocrystalline and coarse grained tungsten: Microstructure, effect of dpa rate, temperature, and grain size. *Acta Mater.* **2018**, *149*, 206–219. [CrossRef]
39. Duan, J.; Wen, H.; He, L.; Sridharan, K.; Hoffman, A.; Arivu, M.; He, X.; Islamgaliev, R.; Valiev, R. Effect of grain size on the irradiation response of grade 91 steel subjected to Fe ion irradiation at 300 °C. *J. Mater. Sci.* **2022**, *57*, 13767–13778. [CrossRef]
40. Miller, M.; Kenik, E.; Russell, K.; Heatherly, L.; Hoelzer, D.; Maziasz, P. Atom probe tomography of nanoscale particles in ODS ferritic alloys. *Mater. Sci. Eng. A* **2003**, *353*, 140–145. [CrossRef]

41. Ukai, S.; Uwaba, T. Swelling rate versus swelling correlation in 20% cold-worked 316 stainless steels. *J. Nucl. Mater.* **2003**, *317*, 93–101. [CrossRef]
42. Wang, X.; Zhang, H.; Baba, T.; Jiang, H.; Liu, C.; Guan, Y.; Elleuch, O.; Kuech, T.; Morgan, D.; Idrobo, J.-C. Radiation-induced segregation in a ceramic. *Nat. Mater.* **2020**, *19*, 992–998. [CrossRef]

MDPI AG
Grosspeteranlage 5
4052 Basel
Switzerland
Tel.: +41 61 683 77 34

Nanomaterials Editorial Office
E-mail: nanomaterials@mdpi.com
www.mdpi.com/journal/nanomaterials

www.ingramcontent.com/pod-product-compliance
Lightning Source LLC
Chambersburg PA
CBHW061627210926
51850CB00009B/2459
9783725875832